STUDIES IN BIOLOGY, ECONOMY AND SOCIETY

General Editor: Robert Chester, Department of Social Policy and Professional Studies, University of Hull

The study of eugenics today has the aim of increasing understanding of our own species and of the rich complexity of the biosocial fabric, so that professional workers, decision-makers in the community and the public at large may be well informed in areas of concern to the whole society. The Eugenics Society promotes and supports interdisciplinary research into the biological, genetic, economic, social and cultural factors relating to human reproduction, development and health in the broadest sense. The Society has a wide range of interdisciplinary interests which include the description and measurement of human qualities, human heredity, the influence of environment and the causes of disease; genetic counselling, the family unit, marriage guidance, birth control, differential fertility, infecundity, artificial insemination, voluntary sterilisation, termination of pregnancy, population problems and migration. As a registered charity, the Society does not act as an advocate of particular political views, but it does seek to foster respect for human variety and to encourage circumstances in which the fullest achievement of individual human potential can be realised.

Amongst its activities the Eugenics Society supports original research via its Stopes Research Fund, co-sponsors the annual Darwin Lecture in Human Biology and the biennial Caradog Jones Lecture, and publishes the quarterly journal *Biology and Society*. In addition, the Society holds each year a two-day symposium in which a topic of current importance is explored from a number of different standpoints, and during which the Galton Lecture is delivered by a distinguished guest. The proceedings of each symposium from 1985 constitute the successive volumes of this series, *Studies in Biology, Economy and Society*. Although the balance between different disciplines varies with the nature of the topic, each volume contains authoritative contributions from diverse biological and social sciences together with an editorial introduction.

Information about the Society, its aims and activities, and earlier symposium proceedings may be obtained from: The General Secretary, The Eugenics Society, 19 Northfields Prospect, Northfields, London, SW18 IPE.

STUDIES IN BIOLOGY, ECONOMY AND SOCIETY

General Editor: Robert Chester, Department of Social Policy and Professional Studies, University of Hull

Published

Milo Keynes, David A. Coleman and Nicholas H. Dimsdale (*editors*)
THE POLITICAL ECONOMY OF HEALTH AND WELFARE

Peter Diggory, Malcolm Potts and Sue Teper (*editors*)
NATURAL HUMAN FERTILITY

Milo Keynes and G. Ainsworth Harrison (*editors*)
EVOLUTIONARY STUDIES: A Centenary Celebration of the Life of Julian Huxley

David Robinson, Alan Maynard and Robert Chester (*editors*)
CONTROLLING LEGAL ADDICTIONS

Series Standing Order

If you would like to receive future titles in this series as they are published, you can make use of our standing order facility. To place a standing order please contact your bookseller or, in case of difficulty, write to us at the address below with your name and address and the name of the series. Please state with which title you wish to begin your standing order. (If you live outside the UK we may not have the rights for your area, in which case we will forward your order to the publisher concerned.)

Standing Order Service, Macmillan Distribution Ltd, Houndmills, Basingstoke, Hampshire, RG21 2XS, England.

Evolutionary Studies

A Centenary Celebration of the Life of Julian Huxley

Proceedings of the twenty-fourth annual symposium of the Eugenics Society, London, 1987

Edited by
Milo Keynes
Department of Anatomy
University of Cambridge

G. Ainsworth Harrison
Department of Biological Anthropology
University of Oxford

MACMILLAN

in association with
Palgrave Macmillan

First published 1989

Published by
THE MACMILLAN PRESS LTD
Houndmills, Basingstoke, Hampshire RG21 2XS
and London
Companies and representatives
throughout the world

British Library Cataloguing in Publication Data
Eugenics Society, Symposium (24th: 1987: London,
England)
Evolutionary studies.
1. Man. Evolution
I. Title   II. Keynes, W. Milo (William Milo)
1924–   III. Harrison, G. A. (Geoffrey
Ainsworth), 1927–   IV. Huxley, Julian   V.
Series
573.2
ISBN 978-1-349-09960-3   ISBN 978-1-349-09958-0 (eBook)
DOI 10.1007/978-1-349-09958-0

Contents

List of Plates

List of Tables

List of Illustrations

Notes on the Contributors

W. H. G. Armytage is Emeritus Professor of Education, University of Sheffield.

Patrick Bateson, FRS, is Provost of King's College, Cambridge, and Professor of Ethology, University of Cambridge.

Bryan C. Clarke, FRS, is Professor of Genetics, University of Nottingham.

M. H. Day was Professor of Anatomy, St Thomas's Hospital Medical School, London, and now works in the Department of Palaentology, British Museum (Natural History).

R. I. M. Dunbar is Lecturer, Department of Anthropology, University College, London.

John R. Durant is Head of Research, Science Museum, and Visiting Professor, Imperial College, London.

G. Ainsworth Harrison is Professor of Biological Anthropology, University of Oxford.

E. B. Ford, FRS (1901–1988) was Professor of Genetics, University of Oxford.

David Hubback is Chairman of the Simon Population Trust.

Sir Andrew Huxley, OM, FRS, is Master of Trinity College, Cambridge, and Past-President of the Royal Society.

T. S. Kemp is Curator of the Zoological Collections and Lecturer in Zoology, University of Oxford.

Robert D. Martin is Director of the Anthropological Institute, University of Zürich-Irchel, Switzerland.

Introductory Note

Sir Francis Galton, FRS (1822–1911), who shared a common grandfather with Charles Darwin (1809–1882) in Dr Erasmus Darwin (1731–1802), formed the idea of eugenics in *Macmillan's Magazine* in 1865, and coined the word *eugenics* in 1883. By this, he wished to mean 'good in birth' and 'noble in heredity'. For him, it denoted the science of improving human stock and allowing the more suitable races, or strains, a better chance of prevailing over the less suitable. He gave the Huxley Lecture in 1901, and started The Galton Laboratory for National Eugenics, or the Eugenics Laboratory, at University College, London, in 1904. In his will, he funded a chair of Eugenics in London University. In 1907, Galton founded the *Eugenics Education Society*, becoming its first President, and on his death he was succeeded by Charles Darwin's fourth son, Leonard Darwin (1850–1943). In 1926, the Society became the *Eugenics Society*, and, in 1989, the *Galton Institute*. The first Galton Lecture of the Eugenics Society was given on Galton in 1914 by Sir Francis Darwin, FRS (1848–1925), the third son of Charles and editor of *The Life and Letters of Charles Darwin* (1887). T. H. Huxley, FRS (1825–1895) could not have been, of course, a Fellow of the Eugenics Society, but his grandson, Julian (1887–1975), was a Life Fellow from 1925. His paper *Eugenics and Eugenists* appeared in 1920, and he gave the Galton Lecture 'Eugenics and Society' in 1936. The Galton Lecture 'Eugenics in Evolutionary Perspective' followed in 1962 and is reprinted in this volume. He was President of the Eugenics Society from 1959 until 1962.

M. K.

1 Introduction

G. Ainsworth Harrison

Julian Huxley's position in the development of evolutionary studies was pivotal. Much of the first half of this century was concerned with accumulating evidence that showed that evolution had occurred, much of the second half (barring the waste of time rebutting the new creationists) has been devoted to analysing how it occurred. Huxley was the outstanding advocate of the 'theory of evolution' in this century, but he was also deeply concerned with identifying evolutionary mechanisms and himself initiated a number of new approaches to unravelling these mechanisms. As well shown here by John Durant's historical review, he thus bridged the two phases and was instrumental in transforming the first into the second.

Central to his commitment to evolutionary theory was his understanding of the power of natural selection. Right at the beginning of his monumental work *Evolution: the Modern Synthesis* (1942). Huxley comments on the strong deductive element in Darwinism. 'Darwin', he says, 'based his theory of natural selection on three observable facts of nature and two deductions from them.' The first two facts were the geometric power of reproductive increase in all species, and the usual constancy of population sizes over considerable time. From these can be deduced the 'struggle for existence'; Darwin's third fact was the phenomenon of variation which is so large that no two individuals can ever be identical. From this fact and the first deduction of a struggle of existence, a second deduction could be made that some variants would have a greater chance of surviving and leaving offspring than others; this is the process of natural selection.*

---

* In the first two editions of *The Origin of Species* (November 1859 and January 1860), Darwin used the phrase 'survival of the adapted', and it is only in the third (April, 1861), and subsequent editions, that he followed: 'I have called this principle, by which each slight variation, if useful, is preserved, by the term Natural Selection . . .' with 'But the expression often used by Mr Herbert Spencer of the Survival of the Fittest is more accurate, and is sometimes equally convenient.' The expression was added, with the exception of sub-headings and the title of the fourth chapter, twice more to the original text, and provided Darwin with a catch-phrase (at least in print), but not necessarily with an improvement. (M.K.)

1

Huxley was a committed selectionist, seeing in differential Darwinian fitness a mechanism which could explain the whole of organic diversity and the endless examples of adaptations which fitted organisms to their environments. He immediately appreciated the ramifying implications of the works of R. A. Fisher, especially *The Genetical Theory of Natural Selection* (1930), and J. B. S. Haldane's *The Causes of Evolution* (1932), which provided the genetic basis for general evolutionary theory and the action of natural selection. He was an avid supporter of E. B. Ford who nicely discusses their collaborative experience in this book, and who was the first to obtain systematic empirical evidence for selection operating in natural populations. And he documented speciation occurring through the gradual change brought about by continuous selection of small heritable differences as envisaged by Darwin himself. Such views were the orthodoxy of the 1940s, manifest not only in Huxley's book but also in the three very influential publications in the United States of T. Dobzhansky's *Genetics and the Origin of Species* (1937), E. Mayr's *Systematics and the Origin of Species* (1942) and G. G. Simpson's *Tempo and Mode in Evolution* (1944). Today that position is very strongly challenged, as thoroughly discussed by Brian Clarke in this volume. Beginning with the pioneering work of Sewall Wright, who recognised that stochastic factors could be extremely important in evolution and developed the concept of genetic drift, the view that much of evolution is due entirely to mutation and chance has received ever increasing support. This view is developed in its most extreme form by M. Kimura in his book *The Neutral Theory of Molecular Evolution* (1983) which argues that the immense variety which has now been uncovered at the biochemical and molecular level does not and cannot have any significant effect on mortality and fertility.

The controversy between 'selectionists' and 'neutralists' is now the central issue in evolutionary biology. Neither side, of course, believes they hold the exclusive truth. Selectionists recognise that chance has played a crucial role in evolution and neutralists accept that organisms show many morphological, physiological and even biochemical characters which are functional adaptations to environments and must therefore be the product of natural selection, but there is enormous disagreement over the extent to which the two processes have contributed to evolutionary change. In favour of the neutralists' view are the facts:

(1) That particular groups of proteins appear to evolve at essentially constant rates over long periods of times.
(2) That much of the variation observable at the DNA level gains no expression in proteins and much of that in proteins appears to have no functional significance.
(3) That if every variable locus carried a fitness differential the segregational load would be impossibly high with such a high proportion of the transcribed genome polymorphic (of the order of 20 per cent in a variety of organisms tested).
(4) That the number of examples where selection has been empirically shown to be operating is trivially small in relation to the magnitude of the variability discovered.

There is a selectionist reply to each of these points. Constancy of molecular evolution may be at least to some extent illusory arising from averaging rates over long periods of time. In considering fitness one needs to take account of the efficiency and economy of productive systems as well as the functional differences in the products. Much of the problem of 'load' disappears if one thinks of fitness as a comparative state dependent upon which other particular genotypes are present in a population rather than in absolute terms. It also needs to be remembered here that selection actually operates on total phenotypes – on whole individuals – not on separate genes as was well recognized by Darwin and Huxley. Finally, detecting selection is a difficult business even when selection pressures are quite high and evidence for selection in a system has usually been obtained when it has been thoroughly and rigorously sought.

Selectionists would also point out that whereas the biochemical and molecular variation may appear unimportant, it is variation at this level which must be responsible for the morphological and physiological variation which most biologists see as subject to selection. Then the fact that all species possess characteristics which seem to fit them to their environment, testifies to selection, even if this cannot be rigorously tested because of the absence of alternative characteristics against which comparison of effect can be made.

The argument is long likely to continue with selection difficult to demonstrate and neutrality impossible to detect. Further, the two processes may well occur in the same system at different times. Thus, for example, a gene which protects against some infectious disease will be selected when the disease is present, but may not when it is absent, and co-evolution of host and parasite can establish host protective systems which become redundant as parasites evolve to overcome them. Neutrality can in fact arise from selection!

Although there is no simple resolution to the issue, it cannot merely be put on one side, since it affects every other aspect of evolutionary debate. In particular it influences thinking about the evolutionary relationship of organisms with one another. 'Similarity indicates phylogenetic affinity' is much more likely to be true under neutral evolution than selection since neutral evolution occurs at constant rate and is less likely to produce convergence.

Evidence for evolution comes from many different sources, but one which had received little attention until Huxley's time was behaviour. This was no doubt because ethology (as distinct from animal psychology) itself had little scientific basis before the brilliant researches of Lorenz, Tinbergen and Von Frisch in mid-century. These workers also tended to concentrate initially more on mechanisms than evolution. Indeed, as recounted in this book by Robin Dunbar some of the very earliest evolutionary approaches to behaviour were actually conducted by Julian Huxley on the great-crested grebe, and the red-throated diver.

Today the position is very different with behavioural studies, especially through sociobiology in centre-stage of evolutionary biology. Neutralists there may be when it comes to consider molecules, but not to behaviour; and ethology provides some of the finest examples of natural selection. The essential difference between the evolutionary study of anatomy, physiology and biochemistry of organisms on the one hand and of behaviour on the other is that the former can be regarded as 'private properties' influencing only the fitness of their possessors, whereas behaviour, and specifically social behaviour, concerns interactions between different individuals and has corporate properties. It is no good having the right message if this cannot be understood and acknowledged. Function may not always be obvious, but it is typically more obvious than a base-pair change.

The roots of sociobiology run deep, from Darwin's interest in sexual selection, through Fisher's ideas on parental investment and Haldane's on kin selection, through to Hamilton's development of the concept of inclusive fitness and Trivers's of reciprocal altruism. Strangely this made little impact in general evolutionary biology until E. O. Wilson pulled together and documented all the strands in his influential work *Sociobiology, the New Synthesis* (1975). Since then sociobiology has become as much a banner for the late twentieth century as Darwinism was for the nineteenth and for the same reason: its implication for the human condition. The fact that animals may use complex and devious behaviours to maximise their reproductive success has been judged to be scientifically interesting if not surprising; the suggestion that the same goal may be at the root of much of human social behaviour is torn with controversy. Sociologists and social anthropologists, almost universally pour scorn on the idea which is perceived as being impudent if not nonsensical! Unfortunately the issue has been confused over the extent of the genetic paradigm, with 'biological' often being confused for 'genetic'.

Few would wish to deny that the human organism has a strong intrinsic drive to reproduce but it is indeed nonsense to suggest that the diversity of mating systems that one finds among human societies represent different genetical programmed adaptations to reproduction in particular ecological circumstances. Evolution is full of the improbable, but there is no way that genetic differentiation of this magnitude and form could have arisen in the short history of *Homo sapiens* and against the endless gene flow between populations. However, human sociobiologists point out that the essence of their model does not require any genetic variance within or between populations and that all diversity is explicable in terms of individual adaptability. The key question is, Do human beings operate to maximise their reproductive success? And do different mating systems represent ways of doing this in different environmental (including social) circumstances? Empirical evidence is now being collected which suggests that the answer to both these questions in non-contracepting societies is *Yes*.

The question of mechanism is a quite separate issue but it has been appreciated now that natural selection can operate on environmentally caused variants as well as genetic ones, and if such variants are transmitted exclusively between parents and offspring a system of change is established which, while not evolution (which requires genetic change) is indistinguishable from it operationally and at the level of the gross phenotype. It would appear that social status tends to be a primary factor in determining reproductive success in humans as in social animals and any factor which enhances status is likely to be advantageous. Social status seems to be a 'universal' in human societies, though in many modern groups it is decoupled from reproductive success by contraception.

Huxley was not a palaeontologist but he was well aware that some of the most important tests for evolutionary theory were in the fossil record. This record testifies magnificently to the fact of evolution, and reveals a host of evolutionary phenomena which could not be expressed elsewhere, like extinction and evolutionary trends. But, as Tom Kemp shows in this book, it is a poor source of knowledge about primary evolutionary mechanism and, especially, the process of speciation. The neo-Darwinist view, promulgated by Simpson and supported by Huxley, is that species arise by the gradual replacement or differentiation of genes under natural selection and genetic drift. This view has been repeatedly challenged, particularly today in the model of 'punctuated equilibrium' supported by Eldredge and Gould and Williamson. This sees evolution progressing by short periods of rapid change followed by long periods with little or no change. The evidence for or against these two conflicting interpretations is finely evaluated by Kemp who points out that the fossil record itself has poor resolving power in the controversy. This, of course, is mainly because it represents only a very small sample in both time and space of what was actually happening. The debate, however, has high-lighted the fact that natural selection operates as often to maintain stasis as to produce change (normalising and stabilising selection versus directional selection), and that some forms have remained unchanged over many millions of years while others have given rise to great adaptive radiations in less time.

It is, perhaps, also worth commenting that this debate follows the general pattern of evolutionary controversy – the 'either/or' one. Protagonists take the view that all speciation must be the result of gradualism or of punctuated equilibrium. From study of living forms, however, we know subspecific variation caused by selection for local adaptations can verge on speciation, and that by cytogenetic change, particularly polyploidy, new variants of specific dimension can arise instantaneously. Perhaps both mechanisms have played a significant role in producing the fossil record!

Huxley was particularly interested in the relationship between the development and evolution of the phenotype: in his introduction to the *Modern Synthesis* he modestly says, 'Any originality which this book [has is in] stressing the fact that a study of the effects of genes during development is as essential for an understanding of evolution as are the study of mutation and that of selection'. The genetic analysis of development has only just begun in most organisms, but an interest in how changes in growth produce changes in form and morphology has been long standing in evolutionary biology. This interest was strongly kindled by Huxley particularly in his book

Problems of Relative Growth (1932). It is also now widely appreciated that selection acts on every stage of the life-cycle: that organisms, throughout the whole of their life history, including all developmental stages, need to be adapted to their environments. Further, at any one time total form has to meet various functional needs, such as feeding, defence and reproduction, which can generate conflicting pressures. Many of these interactions end up as being expressed in body size and shape, and Huxley was greatly interested in the evolution of body size and the allometric relations between the growth of different structures which contribute to whole body and organ size. Modern developments in this area are evaluated by Robert Martin in this volume.

Since he is specially concerned with primates, Martin also introduces the anthropological dimension. Huxley was always interested in the application of general evolutionary principles to human origins and diversity, and he wrote extensively on anthropological affairs. Modern biological dimensions to these are here discussed by Michael Day and myself. But Huxley was well aware that man was not 'just another animal' and that human culture added a totally different dimension to the human experience. Some of this is because culture itself provides innumerable new and diverse environments which must act as selective agents. But much more important is the release culture provides from solely biological processes and the speed with which cultural change, as compared with biological evolution can occur. Further in a sense that has no equivalence in any other organism, human beings can control their own destiny, since they can totally determine natural environments and create cultural environments to their will.

However, as the sociobiological debate has emphasised, mankind has not escaped from all biological constraints, and biological (if not genetic) processes still govern much of our behaviour, including our social behaviour. The understanding of altruism is here of critical concern, since while it can be essentially a social phenomenon it can also be essentially a biological one. This important issue is considered here by Patrick Bateson, particularly in relation to the nature of human ethics.

Huxley's concern for mankind was far from being a solely academic one and he saw understanding of biological evolution as a basis for human action. He became very much a man of public affairs, and two outstanding examples of this were his directorship of UNESCO, a period in his life documented here by Harry Armytage, and his involvement with Eugenics, described in this book by David Hubback. It was, of course, because of the latter and his Presidency of the Eugenics Society, that it seemed so fitting that this Society organised the celebration of the centenary of his birth.

Julian Huxley was certainly one of the outstanding scientists and thinkers of this century: a man whose commitment to understanding, explaining and applying evolutionary principles has affected all our lives. What was it that created his genius? Obviously much that was highly individual, but also a remarkable interplay between genetic and cultural inheritance. Fittingly this inheritance is described in great detail in the first contribution of the book by his half-brother Sir Andrew Huxley, in the Eugenics Society's Galton Lecture for 1987. (The Society became the Galton Institute in 1989.)

2 The Galton Lecture for 1987: Julian Huxley – A Family View

Sir Andrew Huxley, OM, FRS

Julian Huxley was my half-brother. (See Figure 2.1, Family Tree of Julian Huxley 1.) Our father was Leonard Huxley, whose first wife, Julia Arnold, Julian's mother, died young of a cancer in 1908; Leonard married again four years later and I am the younger of the two sons of the second marriage. But although genealogically of the same generation, I was 30 years younger than Julian, so he was more like an uncle to me than a brother. He had left the family home long before I was born, and I was only a year old when he married Juliette; his two sons are only a little younger than myself. The first six years of his married life were spent as a don in Oxford, where he was a teaching fellow of New College and Senior Demonstrator in the Zoology Department; I can just remember one visit to their very attractive home in Holywell. But in 1925 he was appointed to the chair of Zoology at King's College London, and he and Juliette moved to a house on the Holly Lodge Estate in Highgate, not far from our home in Hampstead. From then on, apart from the two years in Paris as Director-General of UNESCO and many shorter periods when he was abroad on his travels, London was his home, and we saw him often.

The best way to get an impression of Julian as a person is to read the magnificent autobiography that his widow published last year (Juliette Huxley, 1986). By comparison, what I can say is very superficial. The phrase that first comes to mind is that it was great *fun* to be in his company. He was interested in almost everything – literature, art, music, history, education, public affairs and most of all animals and plants (but not mechanical things), and would converse on any topic in an animated, articulate and interesting manner. He had an unlimited fund of stories and jokes, often spiced with impropriety. He was extraordinarily good at repartee; the following is an example that my wife and I treasure. During one of Aldous Huxley's last visits from the United States, where he had settled shortly before

9

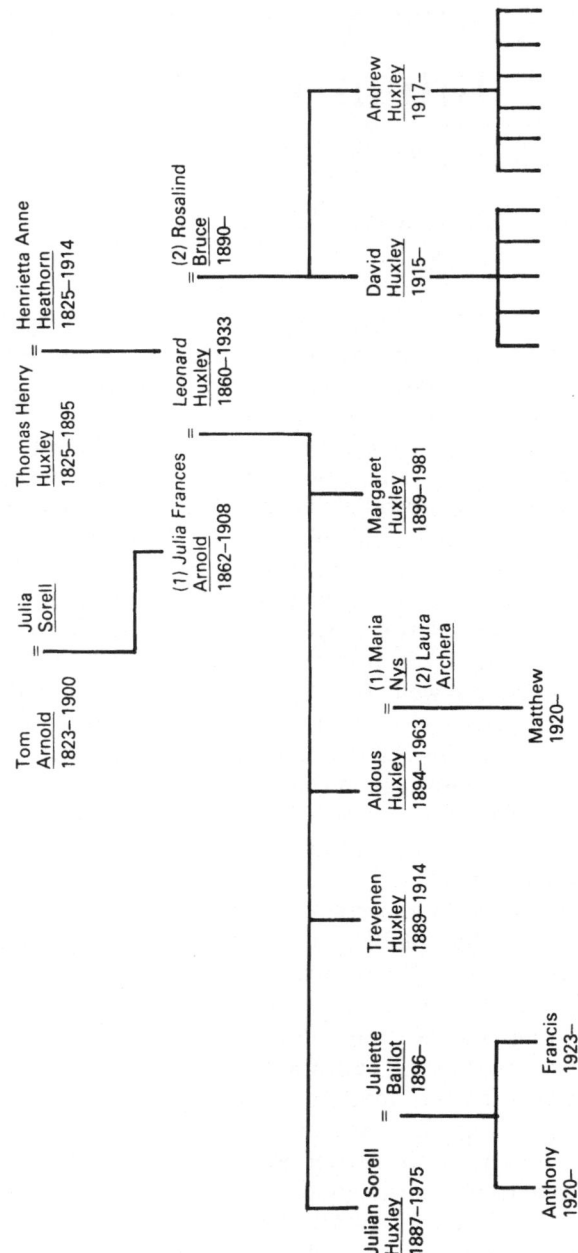

Figure 2.1 Family tree of Julian Huxley 1

the Second World War, Julian and Juliette invited us to dine with them to see Aldous. Aldous remarked that his son Matthew had recently taken a Master's degree in public health and would be able to put the letters MPH after his name, with the implication that it could be misread as standing for 'miles per hour'. But Julian's eyes twinkled, and after only a moment's thought he said: 'What does that stand for – Master of Pox Hounds?'

The other side of the liveliness of his mind was that he was sometimes assertive and impatient, not always thoughtful of others and perhaps too impulsive to be tactful. He admits in his autobiography (*Memories*) to a tendency toward exaggeration when he was young, recording his reply to a rebuke 'for some absurd fabrication': 'Well, of course, the sun's shining and I always *exaggerate* a little when it's a fine day' (J. S. Huxley, 1970, p. 22).

I remember too being told by his cousin Janet, wife of George Trevelyan, that when a large bird of prey was seen during a walk in the Lake District, Julian would identify it as a buzzard if it had been first spotted by some other member of the party, but, if he himself had been first to see it, it was a golden eagle.

He gave the impression of great self-confidence, but in *Memories* he recalls many episodes of self-doubt, and on half a dozen occasions, early and late in his life, these developed into major nervous breakdowns, with complete loss of will-power. I did not see him during any of those attacks, which were evidently very similar to those suffered by our common grandfather, Thomas Henry Huxley. He lived at high pressure and achieved a fantastic amount. The *Biographical Memoir* written for the Royal Society by John Baker (1976), once Julian's pupil, has a select list of almost one hundred books and scientific papers by him, but a much more complete bibliography published in 1978 by UNESCO runs to more than 100 pages, with an average of at least ten items per page (Baker and Green, 1978). Numerically, the majority of these thousand-odd items of published material are articles and letters in newspapers, reprints of broadcasts, etc., but the effort it represents is formidable – the list includes about fifty books (not all of Julian's sole authorship).

Julian's life was lived not only at high pressure but on a high intellectual plane. He concerned himself with the great problems of mankind – spiritual as well as material – and dealt with them both in conversation and in his writings at a level that I was quite unable to match. My feelings are well expressed by a remark made by Sir Joseph Hooker about T. H. Huxley; Hooker, the leading botanist of his time, eight years older than Huxley and President of the Royal Society ten years before him, is recorded in a letter of Charles Darwin as having said, 'When I read Huxley I feel quite infantile in intellect' (L. Huxley, 1903, ii, p. 64). That is how I used to feel in relation to Julian.

There are other personal resemblances between Julian and his Huxley grandfather. I have already mentioned their repeated periods of depression. I spoke also of Julian's lively conversation and fund of stories; most people who know T. H. Huxley only from his portraits and photographs probably think of him as a rather severe and forbidding character but this would be the reverse of the impression he gave both within his family and in social intercourse. T. H. H.'s friend Anton Dohrn, founder of the Marine Station at Naples, used to refer to him and his wife and children specifically as 'the happy family', and there are many accounts of his delightful conversation. Spencer Walpole, who accompanied him as a colleague on many of his journeys as an Inspector of Fisheries, wrote:

> It is needless to say that, as a companion, Professor Huxley was the most delightful of men, . . . He knew how to draw out all that was best in the companion who suited him; and he had equal pleasure in giving and receiving. Our conversation ranged over every subject. We discussed together the grave problems of man and his destiny; we disputed on the minor complications of modern politics; we criticised one another's literary judgments; and we laughed over the stories which we told one another, and of which Professor Huxley had an inexhaustible fund. (L. Huxley, 1903, ii, p. 296)

Published records do not tell whether some of these stories were laced with impropriety in the way that many of Julian's were; I suspect they may have been as I remember my father telling of a scientific meeting at which the speaker gave, as an example of the failure of acquired characteristics to be inherited, the fact that the Jews had practised circumcision for a hundred generations or so without inherited effect, whereupon T. H. H. whispered to his neighbour: 'There's a divinity that shapes our ends, rough-hew them how we will.' Another instance is the sketch reproduced in Nora Barlow's (1958) edition of Charles Darwin's Autobiography: in a letter to Darwin, T. H. H. invents a disease 'Darwinophobia' and says: 'It's a horrid disease and I would kill any son of a [sketch of conspicuously female member of the canine species] I found running loose with it without mercy.'

Both Julian and T. H. H. admit in their autobiographies to having had good opinions of their own abilities when young. Julian quotes a school report by his biology master at Eton*: 'Huxley ma. K.S. Barring a tendency to think himself infallible, I have no possible fault to find with him . . .' (J. S. Huxley, 1970, i, p. 50) while T. H. H. writes of the beginning of his time working at the Naval Hospital at Haslar:

> My official chief at Haslar was a very remarkable person – the late Sir John Richardson . . . and, having a full share of youthful vanity, I was extremely disgusted to find that 'Old John', as we irreverent youngsters called him, took not the slightest notice of my worshipful self, either the first time I attended him . . . or for some weeks afterwards. (T. H. Huxley, 1893, p. 11).

He finished his estimate of his own character, given to Francis Galton in 1873, with the phrase 'disinterestedness arising from an entire want of care for the rewards and honours most men seek, vanity too big to be satisfied by them' (Bibby, 1959, p. xxii).

Julian was descended on both sides from notable nineteenth-century figures. (See Figure 2.2, Family Tree of Julian Huxley 2.) I have said a little about his paternal grandfather, so I will now switch to his mother's side. She was Julia Frances Arnold, a grand-daughter of the famous Dr Thomas Arnold who was responsible as headmaster for raising Rugby, and by example the other English public schools, from the shocking state they had sunk to by the early years of last century. His influence on Julian can only have been remote, as he died young in 1842, but the next generation included five who distinguished themselves in the educational world. Dr Arnold's second son Tom, Julian's grandfather, was a brilliant scholar but led a varied, not to say erratic, life which included periods as a schoolmaster in New Zealand, school inspector in Tasmania, Professor of English Literature in the Catholic University of Dublin, teacher in Cardinal Newman's Catholic school in Birmingham, coach to undergraduates at Oxford, and finally again Professor of English Literature in Dublin, in what had become the Royal University of Ireland. Dr Arnold's eldest son Matthew, famous as poet, held a series of educational posts: briefly a master at Rugby, then secretary to the Marquis of Lansdowne who was responsible as President of the

---

* M. D. Hill, to whom Julian acknowledges that he owed much for his stimulating teaching.

14

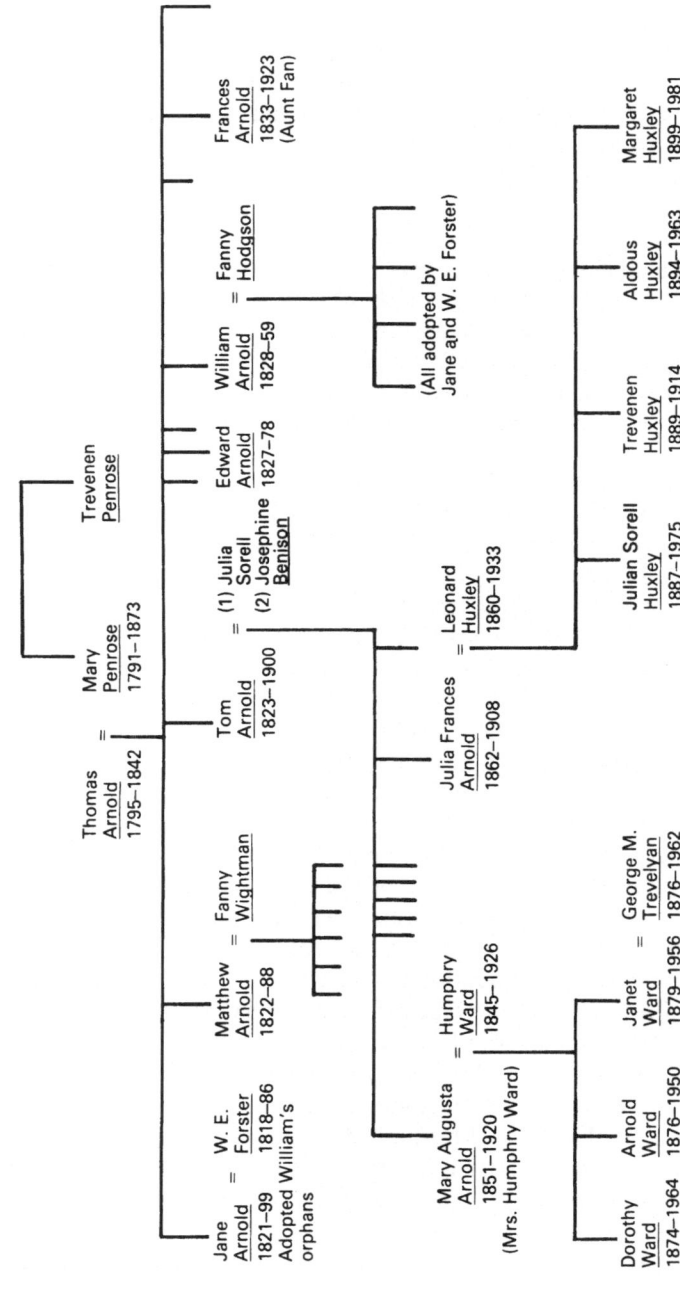

Figure 2.2   Family tree of Julian Huxley 2

Council for the administration of public instruction, and then for the greater part of his life an inspector of schools, with a ten-year period as Professor of Poetry at Oxford.

Tom's next younger brother Edward was also an inspector of schools. Dr Arnold's eldest child was a daughter, Jane; she married W. E. Forster who, as Vice-President of the Council, carried the Endowed Schools Bill and the Elementary Education Bill of 1870. Dr Arnold's younger son William became Director of Public Instruction in the Punjab where he founded over a thousand vernacular schools; he and his wife both died young and their children were adopted by his Forster sister and brother-in-law and took the name Arnold-Forster. Julian's mother, daughter of Tom Arnold, founded Prior's Field school for girls, near Charterhouse, two years after the birth of her youngest child, and was headmistress until her death in 1908; the school still flourishes and keeps a connection with the family, both Julian and myself having been Governors and the present Chairman of Governors being my brother David.

On the Huxley side too, education was a major activity; our father taught classics, briefly at St Andrews University and then at Charterhouse until he joined the publishing firm of Smith, Elder in 1901, while T. H. Huxley's phenomenal activity was divided fairly equally between science, public affairs and education: for most of his life he held posts at the Royal School of Mines and at the Royal College of Science (now Imperial College) into which it developed, largely under his influence; he was an original member – and a dominant one – of the London School Board; and he was an important figure in the development of London University. These and many other activities in the field of education make up the greater part of a full-length book about him by Cyril Bibby (1959).

Both Julian and I benefited directly from reforms for which he was responsible. As a Governor of Eton he had insisted that the school should build science laboratories – known to some as Huxley's Folly – and it was in them that Julian received the first stage of his scientific education. T. H. H. was also the principal figure in getting physiology (my own subject) established as an independent discipline in England: it was on his advice that Trinity College, Cambridge, in 1870 appointed Michael Foster as 'Praelector in Physiology', which was the start of the strong tradition in Cambridge and particularly in my College, Trinity, that attracted me away from physics and into physiology while an undergraduate; and it was he who persuaded T. J. P. Jodrell to offer, and University College London to accept, the benefaction that established in 1874 the Jodrell chair – the first full-time professorship of physiology in England – which I held from 1960 to 1969.

Going one generation further back, T. H. H.'s father George was senior assistant-master at Great Ealing School, well known until it collapsed after the death of its Head Dr Nicholas in 1829.

In so far as Julian's work and interest in education can be attributed to family influences, it can thus be traced fairly evenly to the Arnold and Huxley sides. Much of Julian's life work was indeed educational: as an academic teacher at the Rice Institute (1914–15), at Oxford (1919–25), and at King's College London (1925–27); as populariser of science; and internationally as member of the Colonial Office Committee on Education and finally as Director-General of UNESCO.

Julian's scientific interests, on the other hand, can be traced only to his father's family. It cannot be mere coincidence that both T. H. H. and Julian were first and foremost biologists, both with great emphasis on evolution; in his autobiography Julian says that T. H. H. 'was certainly a great example to me, and his life and writings influenced my own thinking in many fields' (J. S. Huxley, 1970, i, p. 15). On the Arnold side, however, intellectual interests seem to have been exclusively in literature and religion: I have not come across any reference to scientific interests and indeed Matthew Arnold was alarmed by what he regarded as the encroachment of physical science in education, attacking T. H. H. (who, by the way, was a personal friend) unjustly as if the latter's proposals for giving science a share in school education meant the exclusion of literature (Arnold, 1896).

Apart from evolution, however, Julian's interest in biology was very different from T. H. H.'s. At 16, T. H. H. was reading Müller's *Physiology*, and the list of his boyhood interests in his journal *Thoughts and Doings* ranges from metaphysics through mathematics to physical science and also languages but with no reference to any aspect of natural history (L. Huxley, 1903, i, pp. 13–20), while Julian's recollections are full of wild flowers, birds, butterflies and moths (J. S. Huxley, 1970, i, pp. 26–37). In T. H. H.'s short autobiography he describes his interests and ambitions as follows:

> As I grew older, my great desire was to be a mechanical engineer, but the fates were against this and, while very young, I commenced the study of medicine under a medical brother-in-law. But, though the Institute of Mechanical Engineers would certainly not own me, I am not sure that I have not all along been a sort of

mechanical engineer *in partibus infidelium*. I am now occasionally horrified to think how very little I ever knew or cared about medicine as the art of healing. The only part of my professional course which really and deeply interested me was physiology, which is the mechanical engineering of living machines; and, notwithstanding that natural science has been my proper business, I am afraid there is very little of the genuine naturalist in me. I never collected anything, and species work was always a burden to me; what I cared for was the architectural and engineering part of the business, the working out the wonderful unity of plan in the thousands and thousands of diverse living constructions, and the modifications of similar apparatuses to serve diverse ends. (T. H. Huxley, 1893, pp. 6–7)

This description fits my own biological interests precisely, but not Julian's: his interest was in the whole organism, its behaviour, and how it develops. He did spend some months in 1913 in the laboratory of the biochemist Otto Warburg in Heidelberg, but he wrote in his autobiography that he had not been much interested in the biochemical work 'for it largely went through my head, without affecting my research programme: I was always much more interested in the behaviour of living animals and their past evolution than in the physico- chemical basis of their activities'. He went on to Munich to work in Hertwig's laboratory, but 'Here too I failed to profit very much from the work I was put to do – indeed I have forgotten what it was' (J. S. Huxley, 1970, i, pp. 96–7). Dr Durant mentions in his contribution to this book that Julian had been exceptional among evolutionists in continuing to recognise the importance of Darwinian selection throughout the early decades of this century when many Mendelians were assuming that the course of evolution was directed by the occasional occurrence of mutations of large effect; it may be that Julian was saved from falling into this error by his lack of interest in detailed mechanisms. In a similar way, I took up work on muscle with a light microscope about 1953; this had been a profitable approach at the end of the nineteenth century but interest in it was lost as biochemical and biophysical methods became dominant (A. F. Huxley, 1980). I may have been protected from over-emphasis on the biochemical approach by the fact that I never took a formal course in biochemistry.

After his return from the voyage of the *Rattlesnake*, T. H. H.

desired to obtain a Professorship of either Physiology or Comparative Anatomy, and as vacancies occurred I applied, but in vain. . . . At last, in 1854, . . . the Director-General of the Geological Survey offered me the post . . . of Paleontologist and Lecturer on Natural History. I refused the former point blank, and accepted the latter only provisionally, telling Sir Henry that I did not care for fossils, and that I should give up Natural History as soon as I could get a physiological post. But I held the office for thirty-one years, and a large part of my work has been paleontological. (T. H. Huxley, 1893, pp. 14–15)

The great obstacle he had to face was that at that time there were no full-time posts in physiology in England: physiology was taught to medical students by clinicians as a side-line. I have already described the very effective steps that he took to remedy this deficiency. Although he did not carry out any serious research in physiology, he gave many courses of lectures in the subject and wrote a small textbook *Lessons in Elementary Physiology* 'primarily intended to serve the purpose of a textbook for teachers and learners in boys' and girls' schools'; first published in 1866 it was revised three times and went through at least 36 printings altogether.

It is true that T. H. H. devoted himself during the last years of his life to a study of the gentians, their relationships, distribution, hybridisation and evolution. This originated during a visit to Switzerland to convalesce after one of his breakdowns, and developed into a hobby with a serious scientific aspect and also into a love of gardening. He published a 24 page paper on the gentians in the *Journal of the Linnean Society* in 1888. But there was no comparable element of natural history in the scientific work of the main part of his life.

It seems to me that this difference between the scientific interests of T. H. H. and Julian accounts, at least in part, for the difference between the recognition that they received from fellow scientists. T. H. H. was elected to the Royal Society at the very early age of 26, on the strength of papers that he had sent home from HMS *Rattlesnake*, the surveying ship on which he had served as assistant surgeon during a three-year cruise off the coasts of Australia and New Guinea; he had the even more remarkable distinction of receiving the Royal Medal only a year after his election. Julian however was elected at the age of 50, about the average for all FRS's, and received the Darwin Medal when he was nearly 70. T. H. H. was elected on the strength of work in a recognised field, comparative anatomy, in which he had established fundamental relationships between many of the groups of animals among the coelenterates. Julian's work, however, was in some respects ahead of its time. His observations on bird behaviour were in a field not then recognised as a legitimate branch of science; indeed, they formed one of the starting-points of 'ethology'. His experimental work contained many significant contributions toward understanding the processes of development of individual animals, but this is one of the branches of biology in which real understanding has hardly begun to be achieved even now. Perhaps another factor in the late recognition of Julian as a scientist was that he was regarded by his more orthodox colleagues as a 'populariser'; this factor did not arise with T. H. H. as his scientific reputation was established before he began to be known for his approaches to the wider public.

Their activities in bringing biological science and its implications to the lay public, in which both T. H. H. and Julian were quite exceptional among their scientific contemporaries, constitute another parallel between their careers. A large part of the nine volumes of T. H. H.'s *Collected Essays* fall into this category, and he gave huge numbers of lectures outside his formal teaching duties, both discourses to educated audiences at, for instance, the Royal Institution, the British Association and various universities, and his many 'lectures to working men' at the Royal School of Mines, the Working Men's College, the London Institution and elsewhere. He said relatively little about the practical applications of science; the implications he discussed were for the most part philosophical, related to human beliefs, especially of course the impact of Darwinian evolution on Biblical authority and on ideas about 'Man's place in nature' – the title of one of the best known of his essays.

Julian's numerous contributions to the popularisation of science were recognised by the award of the Kalinga Prize in 1953. His greatest single undertaking in this direction was his share in writing *The Science of Life*. This work was conceived by H. G. Wells after the great success of his *Outline of History*, and was written jointly by him, his son G. P. (Gip) Wells, and Julian, Julian taking the largest part in the actual writing. Like the *Outline of History*, it appeared originally in fortnightly parts (31 of them from 1929 to 1930); it was reissued in book form more than once. It was a fairly complete textbook of biology presented in a way that could be appreciated by the intelligent layman without any previous knowledge and was highly successful at that level. Its claim to be intended for the 'ordinary man' was perhaps overambitious but this was not true of Julian's participation in the Brains Trust which was appreciated by a very large section of the population: Julian's contributions not only brought out great numbers of interesting points in biology and their relevance to all sorts of questions, but showed that a man of science can also be remarkably well-informed over a huge range of topics and can have a powerful sense of humour.

As regards the implications of science, at least two of Julian's many other non-technical books, *Religion without Revelation* and *What dare I think*, deal with the same philosophical aspects of science as many of his grandfather's essays, but others range into more practical questions of planning and political issues that T. H. H. did not enter into in his writings. In other respects too, Julian carries the debate further than T. H. H. did; for example, it is interesting to compare their two Romanes lectures, 'Evolution and Ethics' by T. H. H. in 1893 and 'Evolutionary Ethics' by Julian fifty years later (see Professor Bateson's contribution to this book). T. H. H.'s theme was that moral conduct by humans consists in opposing the 'cosmic process', i.e. evolution by the brutal process of natural selection. Julian, however, saw the development of the moral sense in man as itself a part of the evolutionary process; this is an aspect of his division of evolution into a biological phase and a psycho-social phase, only the first of which was considered by T. H. H. It may also be a reflection of Julian's view of evolution as a process leading to overall progress, as emphasised by Dr Durant in his contribution to this book.

Apart from Julian's parents, the members of the Arnold and Huxley families I have so far spoken of died either before or soon after his birth, and any influence they may have had on Julian can have come only through family tradition, through their written works or through family genes. T. H. H. died when Julian was eight years old. Julian records several episodes when they met, and there is a charming exchange of letters between them about water-babies, but I do not suppose their personal contacts can have been of much importance for Julian's later career. His other grandfather, Tom Arnold, lived on until 1900 when Julian was thirteen; he was living in Ireland and had married again after the death of his first wife, Julian's grandmother. The family was also divided on account of Tom's becoming a Roman Catholic, which he did in 1856, only to leave the Church in 1865 because he could not accept the infallibility of the Pope; nevertheless he returned to Catholicism in 1876 and remained a Catholic until his death. Julian in his autobiography does not mention having met him, but he says: 'I inherited my Arnold grandfather's instability of temperament, as well as my Huxley grandfather's determination and dedication to scientific truth' (J. S.

Huxley, preface of *Memories*, 1970, vol. i). There was plenty of temperament also in Julian's Arnold grandmother, born Julia Sorell (the origin of both of Julian's Christian names). She remained a firm Protestant throughout Tom's conversions; Julian's autobiography records that on Tom's first conversion, she 'collected a basket of stones from her yard, walked across to the nearby Chapel where he was being formally received into the ranks of Catholicism, and smashed the windows with this protesting ammunition'(J. S. Huxley, 1970, i, p. 16). She died in the year after Julian's birth.

There were, however, two other members of the Arnold family whom Julian met many times and who must have influenced his development directly. One was Dr Arnold's youngest daughter Frances ('Aunt Fan'), who lived on at Fox How, the holiday home built by Dr Arnold in the Lake District, until her death in 1923. Julian often stayed there as a child, and this acquaintance with the Lake District was one of the sources of his love of nature and the countryside; Julian and Juliette stayed there again on their honeymoon in 1919. The other Arnold I must mention was Julian's Aunt Mary, the eldest child of Tom Arnold and eleven years older than Julian's mother. She married Humphry Ward, an Oxford don who became art critic of *The Times*; she is remembered both as joint founder of the Mary Ward Settlement in Tavistock Place (which still flourishes, though for a different clientele from the slum dwellers for whom it was begun), and as author of *Robert Elsmere, Helbeck of Bannisdale* and 28 other novels, under the name Mrs Humphry Ward. It is said that Julian was portrayed as the child Sandy Grieve in *The History of David Grieve*, written by Mrs Ward when Julian was four years old, partly by the simple device of inserting passages from his mother's letters describing his combination of mischief with irresistible charm (Huws Jones, 1973). Mary Ward was a woman of great intellectual power, and was on familiar terms with many of the important literary and political figures of her time. She must have been a somewhat formidable aunt to Julian, but her country house 'Stocks', near Tring, on the outer edge of the Chilterns was another of the places where Julian often stayed as a boy and developed his love of nature; it was from there that he made his observations of the courtship of the great-crested grebe, which became one of the classics of the scientific study of animal behaviour.

I imagine that Julian's love of nature developed originally through his familiarity with the Surrey countryside. At the time he was born, his (and my) father was a master at Charterhouse and the family lived in the outskirts of Godalming. Julian records that whenever possible, his mother 'would accompany my father and me on our walks, and the two of them did a great deal to instil a love of nature into me, partly Wordsworthian, partly scientific' (J. S. Huxley, 1970, i, p. 20). In the same way, I well remember my father naming the birds and flowers on our Sunday walks on Hampstead Heath, on trips to Surrey and the Chilterns, and on our holidays in Scotland. I am sure I was less receptive than Julian; in those days I never succeeded in recognising birds by their songs, in spite of having them repeatedly named (and imitated) by my father. Although by education and profession a literary man, my father had a good general knowledge of science – physics, chemistry and geology as well as biology; he was of course thoroughly familiar with Darwinian evolution and the controversies that followed the publication of the *Origin of Species*, largely no doubt through being told by his father T. H. H., but also because he had written the *Life and Letters* both of T. H. H. himself and of Joseph Hooker, as well as a short biography of Charles Darwin.

My father would also discuss the local land forms in terms of the underlying geological structures; at the time I did not find this interesting though I now realise how much one's appreciation of hill scenery is enhanced by even an elementary understanding of its structure and history. The origin of words was another topic in which my interest was started by things my father told me during our walks. Probably all these interests go right back to his own childhood: books at our home that he received as school prizes in the 1870s include *Stanley on Birds*, Chenevix Trench's *English Past and Present* and Lardner's *Electric Telegraph*, and a prize he received while at St Andrews before going to Balliol is Helmholtz's *Tonempfindungen*, a major treatise (in German) on the mechanism of hearing and its relation to the theory of music. The mechanism of hearing is one of my own serious interests within physiology; whether this is to be traced to the time when I read this book as an undergraduate, or further back to my father's conversation, I cannot now tell. I am sure that these interests must have come out also in my father's walks with Julian and evidently the natural history aspects struck a deeper chord in him than in me. While a boy I did make collections of butterflies and moths, of grasses, and of ferns, but my interest was mainly a collector's instinct, and I had to rely on my father whenever a difficult distinction between species of grasses or of ferns had to be made by reference to Bentham and Hooker.

Many of the aspects of Julian's temperament that resemble characteristics of his grandparents did not show up in the intervening generation. My father was a very kindly person, never impatient or rude, and with a great capacity for enjoying his many activities. He was entirely stable, with no trace of the fits of depression that affected both his father and Julian, his eldest son. He was an excellent conversationalist, though with less of the sparkle that was characteristic of Julian. I cannot say anything at first hand about Julian's mother; she was evidently a remarkable and very able woman and Julian, like her other children, was devoted to her; he gives a most attractive picture of her in his *Memories* (J. S. Huxley, 1970, i, pp. 18–20). But I have neither read nor heard that she showed any of the instability that Julian believed he had inherited from her parents (preface to *Memories*). It is perhaps worth mentioning that T. H. H. also recollected a deep devotion to his mother (L. Huxley, 1903, i, p. 5). Aldous Huxley was Julian's brother, younger by seven years so that they could not share many boyhood interests, but later friendship developed into an intimacy which was important to both of them. Julian says, 'Conversation with him was a delight, a sharpening of wits with a familiar and companionable mind, which his prodigious knowledge made stimulating and fruitful' (J. S. Huxley, 1970, i, preface), and he gives a fuller appreciation in his second volume (J. S. Huxley, 1973, ii, pp. 95–8).

The love of nature which developed so early and so strongly in Julian led him to become one of the leading figures in nature conservation. He was the first chairman of the Wildlife Conservation Special Committee (England and Wales), whose report in 1947 was the starting-point for the establishment of National Nature Reserves and other governmental action in Britain, under the Nature Conservancy (now the Nature Conservancy Council). One of his great successes as Director-General of UNESCO was the setting up of the International Union for the Conservation of Nature and Natural Resources; later he was active also in the formation of the World Wildlife Fund. His interest in conservation certainly began early in life: he recalls that while at Eton he had to write an essay on 'What would you do if you had a million pounds?' and in his answer he suggested buying up as much as possible of the unspoilt coastline of Britain. The interest in conservation can perhaps be traced back one generation further but not two: Leonard suggested to T. H. H. in 1886 that he should join a society whose object was to prevent a railway being run right through the Lake District, but T. H. H. showed no enthusiasm in his reply:

I am not much inclined to join the 'Lake District Defence Society'. I value natural beauty as much as most people – indeed I value it so much, and think so highly of its influence that I would make beautiful scenery accessible to all the world, if I could. If any engineering or mining work is projected which will really destroy the beauty of the Lakes, I will certainly oppose it, but I am not disposed, as Goschen said, to 'give a blank cheque' to a Defence Society, the force of which is pretty certain to be wielded by the most irrational fanatics among its members . . . People's sense of beauty should be more robust. (L. Huxley, 1903, ii, p. 454)

The change of attitudes over a century no doubt reflects the damage that has been done to the British countryside during that period.

I have sketched some of the resemblances and differences between Julian and various members of his family. No doubt some of the similarities are due to biological inheritance, but it would be idle for me to attempt to unravel the relative contributions made in this instance by nature and by nurture.

Acknowledgement

Much of my information about the Arnold family is derived from Meriol Trevor's book *The Arnolds* (London: The Bodley Head, 1973).

References

Arnold, M. (1896) *Discourses in America* (London: Macmillan) (reprinted 1970 by Scholarly Press, St Clair Shores, Michigan, USA).

Baker, J. R. (1976) *Julian Sorell Huxley*. Biographical Memoirs of Fellows of the Royal Society, *22*, 207–238.

Baker, J. R. and Green, J.-P. (1978) *Julian Huxley, Scientist and World Citizen, 1887 to 1975* (Paris: UNESCO).

Barlow, N. (1958) *The Autobiography of Charles Darwin* (London: Collins) p.211.

Bibby, C. (1959) *T. H. Huxley, Scientist, Humanist and Educator* (London: Watts).

Huws Jones, E. (1973) *Mrs Humphry Ward* (London: Heinemann) p.96.

Huxley, A. F. (1980) *Reflections on Muscle* (Liverpool: University Press) p.19.

Huxley, J. S. (1970, vol. i; 1973, vol. ii) *Memories* (London: George Allen & Unwin).

Huxley, Juliette (1986) *Leaves of the Tulip Tree* (London: John Murray).
Huxley, L. (1903) *Life and Letters of Thomas Henry Huxley* (3 vols) (London: Macmillan).
Huxley, T. H. (1893) *Autobiography*. In *Collected Essays*, vol. 1 (London: Macmillan). Also reprinted in: de Beer, G. (1974) *Charles Darwin, Thomas Henry Huxley, Autobiographies* (London, New York and Toronto: Oxford University Press).

3 Julian Huxley and the Development of Evolutionary Studies

John R. Durant

More than any other British biologist, except perhaps his own paternal grandfather Thomas Henry, Julian Huxley merits the epithet: Statesman of Science (this was the title chosen for the other Huxley Centenary symposium, which was held at Rice University, Texas just a few days after the Eugenics Society meeting in London). Throughout the middle half of this century, Julian Huxley occupied a very special place in the British scientific community. He was, of course, a major biological investigator in his own right; but in addition, he exerted a considerable influence over his colleagues as a teacher, talent-spotter, research collaborator, theoretical trend-setter, organiser and, in later life, public figure-head. And as if all this were not enough, throughout his long career he played a key part in promoting the cultural importance of science before policy-makers, politicians and the general public, both nationally and internationally.

The task of summarising Huxley's biological achievements, let alone assessing them, is daunting. A generalist in an age of increasing specialisation, he worked in half a dozen different fields and kept himself well informed in half a dozen more. The historian Frederick Churchill (1980) has well described him as 'an eclectic of extraordinary range'; but though this captures the breadth of Huxley's work, it scarcely does justice to its depth. For the truth of the matter is that, from a remarkably early age, Huxley acquired a mastery of his subject that enabled him to move through it with great assurance, assessing progress in disparate fields and perceiving significant relationships where others saw only a myriad of particular and unrelated findings. In this role as grand strategist, Huxley worked consistently for the unification of biology; it is no accident that it was he who dubbed the renaissance of Darwinian evolutionary theory in the 1930s and 1940s, 'The Modern Synthesis'.

HUXLEY THE EVOLUTIONIST

Evolution was, of course, the key to all of Huxley's intellectual endeavours; it was the theme that gave coherence to his huge output of research papers, articles, edited collections, monographs, synthetic volumes, textbooks and popular writings. In his autobiography, Huxley (1972) tells of his attendance at the centenary celebrations of Darwin's birth at Cambridge in 1909. On that occasion, his thoughts had gone back to the historic battle between his grandfather and the Bishop of Oxford, Samuel Wilberforce. He had realised more fully than ever, he wrote, 'that Darwin's theory of evolution by natural selection had emerged as one of the great liberating concepts of science, freeing man from cramping myths and dogma, achieving for life the same sort of illuminating synthesis that Newton had provided for inanimate nature . . .'. 'I resolved', he went on, 'that all my scientific studies would be undertaken in a Darwinian spirit and that my major work would be concerned with evolution, in nature and in man.'

In later life, Huxley lived up fully to this early resolution. Time and again, he brought the results of particular projects to bear on matters Darwinian: field-studies of courtship in birds over a period of more than twenty years were conducted with a view to clarifying the roles of natural selection and sexual selection in the evolution of behavioural displays; experimental investigations with his student E. B. Ford on the genetics of the amphipod *Gammarus chevreuxi* were used to illustrate the evolutionary importance of rate-genes; quantitative analysis of relative growth rates confirmed the evolutionary importance of rate-genes, and at the same time cast doubt on the validity both of Ernst Haeckel's so-called Law of Recapitulation and of Henry Fairfield Osborn's principle of Orthogenesis. Almost wherever he turned, Huxley found materials for the study of evolution.

For Huxley, evolution was no mere biological fact; rather, it was a central philosophical principle. In tandem with his biological research, Huxley developed a highly distinctive evolutionary philosophy of nature and society. At the heart of this philosophy was the conviction that evolution, though purposeless, was both materially and morally progressive. Darwinism, Huxley once wrote (1937a), 'has at last given man that assurance for which through all his recorded days he has been searching. It has given him the assurance that there exists outside of himself a "power that makes for righteousness".' On the basis of this claim, Huxley opposed his grandfather's ethical stoicism in favour of a form of evolutionary ethics (Huxley, 1942); and in his more general and popular writings (e.g. 1923, 1964) he advocated this evolutionary ethics as the basis for a new secular religion of scientific humanism. I shall return to the relationship between Huxley's Darwinism and his evolutionary humanism.

THE EVOLUTIONARY SCENE, 1900–1920

Before assessing Huxley's particular contributions as an evolutionary biologist, it is necessary to review the state of evolutionary theory in the first two decades of this century. This period, it will be recalled, was a low-point in the fortunes of Darwinism. To be sure, all serious biologists were by now convinced of the fact of evolution; but only a minority of them shared Darwin's view that natural selection was its principal cause. Instead, the majority opted for a variety of anti-Darwinian theories. There was Mutationism, the belief that single-step genetic changes of very large effect were of primary importance; there was Orthogenesis, the idea that internally directed, non-adaptive trends were the key; and there were several different varieties of Neo-Lamarckism, which held in common a commitment to the central role of the inheritance of acquired characters.

The reasons for what Julian Huxley himself once referred to as 'the eclipse of Darwinism' (Huxley, 1942) at this time are complex. Certainly, natural selection faced several major technical obstacles in the late-nineteenth century, not the least of which was the lack of an adequate underpinning in genetics; but in addition, it faced major philosophical and religious obstacles as well. As the historian Peter Bowler (1983) has shown, many late-Victorian biologists were looking for a teleological or even an explicitly theistic account of evolution, in preference to the non-teleological and frankly materialistic account of the Darwinists. For a considerable period, scientific and ideological factors conspired together against against a Darwinian view of evolution.

Surprising as it now seems, the rediscovery of Mendelism around 1900 served at first only to make matters worse for Darwinism. For in the bitter conflict between the first generation of Mendelians and their opponents, the Biometricians, the Mendelian view of particulate (and hence, discontinuous) inheritance came to be seen as diametrically opposed to the Darwinian view of evolution as gradual (and hence, continuous). In other words, Mendelism was equated with Mutationism. In retrospect, of course, it is easy for us to spot the mistake involved here (Mendelism is a discontinuous theory of inheritance, not a discontinuous theory of the origin of new species); but this is the wisdom of hindsight. At the time, the conflation of Mendelism with an anti-selectionist view of evolution caused great confusion; indeed, it split the biological world in a way which was to delay the establishment of a coherent theory of evolutionary change for perhaps a generation.

That this was so is evident from the initial lack of impact of the population geneticists Ronald Fisher, J. B. S. Haldane and Sewall Wright. It is well known, of course, that in the late-1920s and early-1930s these men laid the foundation of the modern synthesis by integrating Mendelism with the theory of natural selection. What is not so widely appreciated, however, is the extent to which these efforts at first failed to heal the rift between the principally laboratory and study-based geneticists, on the one hand, and the principally museum- and field-based Darwinists, on the other. As Ernst Mayr (1980) has pointed out, in the 1920s and early-1930s most Darwinists simply did not realise the extent to which the population geneticists had overcome the anti-selectionist orientation of turn-of-the-century genetics. As a result, these Darwinists continued their evolutionary studies without the benefit of any firm theoretical basis for their work.

HUXLEY'S DARWINISM

This profoundly unsatisfactory situation provides the context for an assessment of Huxley's early contributions to evolutionary theory. Remarkably, Huxley was never greatly impressed by any of the anti-Darwinian traditions that flourished in the two decades after 1900. There are several possible reasons why this was the case: first, his family background, and particularly the example of his grandfather – known to many late-Victorians as 'Darwin's Bulldog'; second, his schooling at Eton and early training at Oxford under a succession of committed Darwinian biologists – M. D. Hill, Ray Lankester, E. S. Goodrich, J. W. Jenkinson and Geoffrey Smith; and third, his early interest in scientific bird-watching, which brought him under the influence of several enthusiastically Darwinian field naturalists – notably, the amateur ornithologists Edmund Selous and H. Eliot Howard (for more detailed accounts of Huxley's debt to these ornithologists, see Durant, 1981; Burkhardt, in press).

Whatever may have been the relative importance of these influences, their combined effect is clear. From the outset, Huxley's outlook was overtly and self- consciously Darwinian. The earliest paper of Huxley's that I have seen – a manuscript of a talk on 'Habits of Birds' which was read to the Decalogue Club at Balliol College, Oxford in 1907 – demonstrates this fact very clearly (Huxley, 1907). There is nothing in it of dramatic De-Vriesian mutations, of enigmatic orthogenetic trends, or of wistful Neo-Lamarckian strivings. 'An instinct', Huxley told his undergraduate audience, 'is the result of the summation of accidental tendencies, and this summation is brought about by Natural Selection'. Darwin himself might have written such a statement.

BIRD COURTSHIP AND SEXUAL SELECTION

By looking at just two of the many strands of Huxley's research in the first half of his career, it is possible to see how he contributed to Darwinian evolutionary theory in the period just before the emergence of the modern synthesis. First, consider his work on bird courtship. In the spring of 1912, Huxley spent a short holiday watching great crested grebes (*Podiceps cristatus*) on the reservoirs at Tring in Hertfordshire. Males and females were similar in appearance, and they engaged in an extraordinary variety of strikingly beautiful courtship ceremonies (Figure 3.1). Three things about these ceremonies intrigued Huxley (1914): first, they were mutual (i.e. the same ceremonies were performed identically by both male and female); second, they were highly stereotyped, with particular movements being repeated over and again; and third, they were 'self-exhausting' (i.e. they were performed for their own sake, and not as a prelude to mating).

At first Huxley took these displays to be the products of Darwinian sexual selection by female choice; but soon he decided that this could not be so. For not only were the displays performed identically by both sexes, but also they continued to occur long after mate choice had taken place. Instead of sexual selection, therefore, Huxley proposed that the courtship ceremonies were the result of 'mutual

Figure 3.1 Courtship habits of *Podiceps cristatus*. From J. S. Huxley (1914) 'The Courtship Habits of the Great Crested Grebe (*Podiceps cristatus*); with an addition to the Theory of Sexual Selection', *Proc.Zool.Soc.London*, Vol. 2, pp. 491–562, plates I & II.

selection', and he suggested that they served the function of keeping paired birds together. Strong pair bonds, he reasoned, were essential in a species where both parents were needed to protect and care for the young. In this case, such bonds were maintained by a series of elaborate mutual displays, the elements of which had evolved by a process of 'ritualization' in which simpler behavioural elements serving one function had been gradually elaborated and stereotyped in the service of another.

Huxley continued to publish observational studies of bird courtship for a period of around 20 years. An interesting feature of these studies is that consistently – indeed, increasingly as the years went by – they down-played the role and importance of true Darwinian sexual selection by female choice. Thus, Huxley's last major reviews of sexual selection (Huxley, 1938a; 1938b) were real dampeners. Given that among the leading Darwinians Huxley was generally recognised as an authority on this evolutionary mechanism, his judgement concerning its relative unimportance may help to explain the almost total absence of the theory of sexual selection from the modern synthesis.

Partly, Huxley's downplaying of Darwinian sexual selection reflected his persistent preference for working with monogamous bird species in which the sexes are similar in both appearance and courtship behaviour. (Frankel (1988) has suggested that this preference may have arisen out of Huxley's personal distaste for the human institution of polygamy.) Darwin, of course, had always regarded the theory of sexual selection by female choice as applying especially to polygamous species in which males are distinguished from females by elaborate ornamentation and/or display; for him, monogamous species in which the sexes look and act alike were poor material for the study of female choice. In addition, however, I believe that Huxley's lack of enthusiasm for Darwin's second selective mechanism may have reflected his aversion to the idea that caprice on the part of females might carry the evolutionary process in directions that were both unpredictable and, very possibly, undesirable from the point of view of the species (Durant, in press).

With time, Huxley's early paper on the courtship habits of the great crested grebe came to be seen as a landmark in the history of animal behaviour study. In 1966, Huxley was invited by the Royal Society of London to convene a meeting on 'Ritualization of Behaviour in Animals and Man'; and two years later, the grebe paper was reprinted in book form with a foreword by Desmond Morris (Huxley, 1968). In the years after 1914, however, the grebe paper does not appear to have made any very great impact. In fact, it was not until

Huxley's work came to the attention of Konrad Lorenz and Niko Tinbergen in the 1920s and 1930s that it began to bear fruit. Very rapidly, these men incorporated Huxley's ideas on the nature and significance of courtship 'rituals' into their own more general and more theoretically ambitious approach to animal behaviour study. Thereafter, of course, the subject made rapid progress; and in 1973 Lorenz, Tinbergen and Karl von Frisch were awarded a Nobel Prize in recognition of their role in the founding of ethology.

RELATIVE GROWTH AND NATURAL SELECTION

At the same time that Huxley was pursuing his observational studies of bird courtship in the field, he was also engaged in a wide range of more obviously professional biological studies in the laboratory. Huxley began his laboratory studies shortly after his graduation in 1909, when he spent a year at the Naples Zoological Station studying regeneration in the sponge *Sycon*. However, his career as an experimentalist did not really blossom until he returned to Oxford after the First World War. Here during the 1920s, he and his students engaged in a wide range of laboratory studies on differentiation, metamorphosis, physiological genetics and relative growth.

Of these studies, it is arguable that those on relative growth were the most important. Beginning in the early-1920s, Huxley undertook a systematic analysis of what he termed heterogony, or differential growth. He showed that the simple case of the constant differential growth of a part, y, in relation to the rest of the body, x, could be represented by the formula: $y = bx^k$, where b is a constant and k is a measure of the differential growth of the part in relation to the whole. In trickier cases – as, for example, when the parts in question are antlers, and thus are periodically shed – Huxley showed that equally simple algebraic rules applied. In his view, these and similar findings provided clues to the nature of the developmental-genetic mechanisms governing relative growth, both in individuals and in groups of related species (Huxley, 1924).

Huxley believed that heterogony (which he later renamed allometry) had important evolutionary implications. For example, comparisons between groups of related species had often revealed steady trends in, for example, antler or horn size. Unable to think of any adaptive significance in such trends, many palaeontologists had referred them to internal, 'orthogenetic' principles impelling lineages in particular directions. Huxley proposed an alternative solution.

Imagine a hypothetical population possessing a developmental-genetic mechanism governing the relative growth of body- and antler-size. If this population were to be selected for increased body size alone, its descendants would naturally come to possess larger and larger antlers; and this, even though selection was not operating directly on antler size at all.

As Stephen Jay Gould (1980) has observed, this argument works equally well the other way round; that is, selection for increased antler size may also generate corresponding increases in body size. A choice between these alternatives lies partly in a judgement about which (if any) of the two characters is non-adaptive. Either way, however, allometry can explain apparently non-adaptive trends by interpreting them as correlates of other, presumptively adaptive trends.

There is a potential problem with this argument, namely that the developmental-genetic systems generating allometric relationships may themselves be subject to selection. If the famous Irish Elk stood to benefit from having a bigger body but not from having bigger antlers, why was not the allometric relationship between the two altered? Huxley recognised this problem, but still maintained his ground by invoking a secondary assumption, namely that the developmental-genetic mechanisms governing allometric relationships were single and simple, and thus difficult or even impossible to change by selection. Granted this secondary assumption, then, as he put it, the explanatory 'burden' placed upon natural selection by the existence of apparently non-adaptive evolutionary trends was lightened considerably (Huxley, 1932).

HUXLEY AND THE 'MODERN SYNTHESIS'

Though substantially different in content, Huxley's studies on bird courtship and relative growth were formally similar in one sense; for each attempted to make theoretical progress by bringing together branches of biological investigation that were almost entirely separate from one another. In the case of bird courtship, Huxley self-consciously attempted to unify the perspectives of the field naturalist, the professional zoologist and the psychologist (Durant, 1986). In the case of relative growth, Huxley was more squarely on the ground of the professional zoologist; but even here, his work entailed the ambitious attempt to bring together the separate fields of developmental and evolutionary biology in a single discipline. Each of these efforts at synthesis was relatively successful; but it was a third and altogether more ambitious effort, made in the period 1936–42, that was to prove the most successful of all.

In 1936, Huxley acted as President of the Zoology Section of the British Association for the Advancement of Science. In his Presidential Address, he called for the reunification of biology around the core principles of Darwinian evolutionary theory. Significantly, Huxley's address was entitled 'Natural Selection and Evolutionary Progress', and its major points were as follows: first, mutation (in the modern, not the De-Vriesian 'Mutationist' sense), recombination and selection together constituted the principal mechanism of evolutionary change; second, enough was now known about this mechanism to facilitate its fruitful application, not only to small-scale evolutionary phenomena, but also to higher-order phenomena such as adaptation, speciation, extinction and long-term evolutionary trends; and third, such application revealed that the evolutionary process, whilst being slow, cruel and, from a metaphysical point of view, purposeless, was none the less in the long run progressive in a way that might be made to serve as a guide in formulating humanity's goals (Huxley, 1936).

These were to be the enduring themes of all Huxley's later writings on the modern synthesis. Understandably enough, they attracted a great deal of interest, and before long Huxley found himself being urged by friends and colleagues to extend his arguments in the form of a book. Thus it was that he set to work on what became *Evolution: The Modern Synthesis*, which was published six years later. Enormously expanded (it runs to nearly 600 pages of text), and containing exhaustive reviews of all the relevant literature, this work was undoubtedly Huxley's greatest contribution to evolutionary biology. Appearing almost simultaneously with Ernst Mayr's equally authoritative, though shorter, work, *Systematics and the Origin of Species* (Mayr, 1942), Huxley's book served notice that evolutionary biology was entering a new era; an era in which a theoretically reinvigorated Darwinism would be applied to extinct and living organisms in order to gain a better understanding of their complexity, diversity, distribution and dynamic interactions.

Huxley was one of around half a dozen biologists who helped to fashion the modern synthesis in the period 1936–48. This being the case, it is perhaps worth noting one or two distinctive features of his contribution. First, Huxley's was a pluralistic synthesis. Though he would accept no compromise with the creationists, the teleologists or the Lamarckians, in most other respects Huxley was prepared to accommodate a variety of mechanisms as having a part to play in the evolutionary process. For example, he accepted the reality of both adaptive and non-adaptive characters; and by the same token, he acknowledged the role of both selective and accidental factors as causal agencies. Similarly, he gave a place to both selection between individuals and selection between groups. Typical of Huxley's pluralism was his treatment of speciation: 'if Darwin were writing today', he suggested, 'he would call his great book *The Origins*, not *The Origin of Species*' (Huxley, 1942).

In these and similar ways, Huxley conformed well to at least one part of Stephen Jay Gould's historical thesis about the 'hardening' of the modern synthesis. According to Gould (1983), the founders of the modern synthesis originally took a rather 'soft', i.e., pluralist view of the nature and causes of evolutionary change; but with time, he suggests, their position hardened into a far more rigid insistence on the all-pervasiveness and all-sufficiency of natural selection. Huxley certainly began soft. With time, also, he (like virtually all other Darwinists in the post-war period) certainly became more confident about the power of natural selection to effect change. However, I am rather doubtful that his views ever hardened into quite the rigid form that Gould detects in the later work of several of the other founders of the modern synthesis.

A second and equally striking feature of Huxley's synthesis was that from the outset it was concerned with the relationship between Darwinism and progress. The very title of Huxley's 1936 British Association address reveals this concern; and it is also reflected in the structure of *Evolution: The Modern Synthesis*, the title of whose last chapter is 'Evolutionary Progress'. Huxley's preoccupation with the idea of progress reflected the coming together of his scientific beliefs, which as we have seen were fundamentally Darwinian, and his ideological commitment to a form of humanism that looked to evolutionary biology for new moral and political guidelines to replace the discredited claims of conventional religion. In seeking to be a consistent Darwinian evolutionary humanist, Huxley was faced with an acute dilemma. For as a Darwinian, he was committed to the principle that evolution was essentially purposeless; but as an evolutionary humanist, he was equally committed to the principle that evolution was essentially progressive. The question was: how could both of these things be true?

It is arguable that this question constituted the central philosophical problem of Huxley's life-work. The tension between Darwinian purposelessness and humanistic progress is frequently just below the surface, even in his most technical biological writings. Take, for example, *Evolution: The Modern Synthesis*. Despite its title, this book operates with two quite different and essentially unharmonised conceptions of evolution: the one is basically a population-genetic conception drawn principally from work on small-scale genetic change in idealised populations of Mendelian characters; but the other is essentially a progressivist conception drawn principally from a centuries-old European tradition of conceptualising the living world as what Arthur Lovejoy (1936) termed 'The Great Chain of Being'.

The first half of *Evolution: The Modern Synthesis* operates chiefly with the first of these conceptions; but as Huxley turns his attention to larger-scale evolutionary phenomena, so the second conception begins to assert itself. A key change occurs over halfway through the book when, quite without warning, Huxley suddenly announces that, 'Evolution may be regarded as the process by which the utilization of the earth's resources by living matter is rendered progressively more efficient'. Interestingly, this new definition is immediately followed by an attempt to distinguish sharply between short-term evolutionary changes responsible for systematic differences at or below the species level and long-term evolutionary changes responsible for higher-level systematic diversity. 'Much of the minor systematic diversity to be observed in nature', Huxley writes, 'is irrelevant to the main course of evolution, a mere frill of variety superimposed upon its broad pattern'.

The key point to notice here is that the population-genetic and the progressivist conceptions of evolution are by no means synonymous. On the contrary, it is not at all clear that they are even mutually consistent. For as Huxley well knew, the whole point of the population-genetic conception was that it permitted the evolution of all sorts of characters – more 'efficient' and less 'efficient', 'progressive' and 'retrogressive', 'good' and 'bad' – according to the single Darwinian criterion of relative reproductive success. By contrast, the progressivist conception singled out just one type of character by means of the non-Darwinian criterion of absolute functional efficiency. (Elsewhere, Huxley added other absolutist non-Darwinian criteria, including level of organisation, capacity for self-regulation, and capacity for further advance.) But on what grounds was the Darwinist to single out just one among the myriad kinds of genetic change produced by natural selection and accord this privileged status, either as representative of a supposed 'overall direction', or even as definitive of the very process of evolution itself?

Time and again throughout his career, Huxley wrestled with this question. This was the real issue at stake, for example, in his disagreement with Theodosius Dobzhansky in the 1960s about the definition of Darwinian fitness. Huxley objected to Dobzhansky's population-genetic definition of fitness in terms of relative reproductive success. Such 'geneticism', Huxley argued, ignored the progressive phenotypic qualities by virtue of which such relative reproductive success was achieved. In his reply to this charge, Dobzhansky noted that Huxley's definition of fitness in terms of biological improvement lacked rigour; but he went on to make an even more telling point. While natural selection did sometimes produce biological improvements, he wrote, 'I shall nevertheless persist in my "geneticism", and maintain that natural selection does not always or necessarily yield these things' (Dobzhansky, 1963). Dobzhansky had put his finger on the crucial point that from a Darwinian point of view, Huxley's progressivist criterion of fitness (and with it his entire philosophy of evolutionary humanism) was simply incoherent.

For a period of around half a century, Darwinian purposelessness and humanistic progression remained in tension at the heart of Huxley's evolutionary philosophy. In some places, as in his attempt in 1942 to draw a sharp distinction between levels of evolutionary change, this tension threatened the very coherence of his 'synthetic' theory; in others, as in his endorsement of the mystical and fundamentally un-Darwinian evolutionary philosophy of the French Jesuit Teilhard de Chardin (Huxley, 1959), it led him to form alliances that certainly surprised and probably disturbed many of his more orthodox colleagues. In reality, though, no one should have been particularly surprised. For Huxley was never simply a Darwinian biologist. Rather, he was a philosopher of evolution on the grand scale. His vision of the progressive development of nature and society was arguably the nearest thing we have had in the twentieth century to the nineteenth-century evolutionary philosopher Herbert Spencer's so-called 'Synthetic Philosophy'.

CONCLUSION

In the course of this essay I have commented critically on particular aspects of Huxley's thought. Such criticisms are easily made with the wisdom of hindsight; but carried too far, they become little more than an occasion for indulgent self-congratulation. In conclusion, therefore, it may be worth re-emphasising the scale of Huxley's achievement as an evolutionary biologist. Julian Huxley will always be remembered as one of the principal craftsmen of the modern evolutionary synthesis, that most creative combination of population genetics, ecology, biogeography, systematics and palaeontology. Like most craftsmen, Huxley worked with materials many of which were not of his own making; but in bringing them together, he helped to fashion something that was undoubtedly a, great deal more than the sum of its parts.

In the preface to the first volume of his memoirs, Huxley observed: 'I have been accused of dissipating my energies in too many directions, yet it was assuredly this diversity of interests which made me what I am'. In this essay, I have suggested that it was not only Huxley's diversity of interests but also his quest for intellectual unity among them that contributed to his greatest scientific successes. Huxley once penned a popular essay on the history of science (Huxley, 1937b). In it, he wrote the following words:

The great body of ascertained fact which exists in the records of natural science is only a vast stamp- collection, no more than a lumber-room, unless each generation in its turn will make it live. It lives most strongly . . . by being woven into the . . . background of some general philosophy of things.

Huxley did not intend it, of course, but this comment serves as a fitting epitaph to his own life-work in evolutionary biology.

References

Bowler, P. J. (1983) *The Eclipse of Darwinism. Anti-Darwinian Evolution Theories in the Decades Around 1900* (Baltimore & London: Johns Hopkins University Press).

Burkhardt, R. W. (in press) 'Huxley and the Rise of Ethology', in A. Van Helden (ed.), *Julian Huxley, Statesman of Science* (Houston: Rice University Press).

Churchill, F. B. (1980) 'The Modern Evolutionary Synthesis and the Biogenetic Law', in E. Mayr and W. B. Provine (eds), *The Evolutionary Synthesis: Perspectives on the Unification of Biology* (Cambridge, Mass: Harvard University Press).

Dobzhansky, T. (1963) Letter to Julian Huxley, 14 March 1963. In *The Huxley Papers*, Rice University Texas, Box 34.

Durant, J. R. (1981) 'Innate Character in Animals and Man: A Perspective on the Origins of Ethology', in C. Webster (ed.) *Biology, Medicine and Society, 1840–1940* (Cambridge, London and New York: Cambridge University Press).

Durant, J. R. (1986) 'The Making of Ethology: the Association for the Study of Animal Behaviour, 1936–1986', *Animal Behaviour*, vol. 34, pp. 1601–16.

Durant, J. R. (in press) 'The Tension at the Heart of Huxley's Evolutionary Ethology', in A. Van Helden (ed.), *Julian Huxley: Statesman of Science* (Houston: Rice University Press).

Frankel, S. (1988) 'Julian Huxley's Early Work on Sexual Selection in Birds', unpublished manuscript.

Gould, S. J. (1980) 'The Misnamed, Mistreated, and Misunderstood Irish Elk', in S. J. Gould, *Ever Since Darwin* (Harmondsworth: Penguin).

Gould, S. J. (1983) 'The Hardening of the Modern Synthesis', in M. Grene (ed.), *Dimensions of Darwinism: Themes and Counter-Themes in Twentieth Century Evolutionary Theory* (Cambridge, London, and New York: Cambridge University Press).

Huxley, J. S. (1907) 'Habits of Birds' (manuscript). *The Huxley Archive* (Oxford: Edward Grey Institute of Field Ornithology).

Huxley, J. S. (1914) 'The Courtship Habits of the Great Crested Grebe (*Podiceps cristatus*); with an addition to the Theory of Sexual Selection', *Proc. Zool. Soc. London*, vol. 2, pp. 491–562.

Huxley, J. S. (1923) *Essays of a Biologist* (London: Chatto & Windus).

Huxley, J. S. (1924) 'Constant Differential Growth Rates and Their Significance', *Nature*, vol. 114, pp. 895–6.

Huxley, J. S. (1932) *Problems of Relative Growth* (London: Methuen).

Huxley, J. S. (1936) 'Natural Selection and Evolutionary Progress', *Report of the British Association for the Advancement of Science*, vol. 106, pp. 81–100.

Huxley, J. S. (1937a) 'Evolution and Purpose', in J. S. Huxley, *Essays in Popular Science* (London: Pelican Books).

Huxley, J. S. (1937b) 'On the History of Science', in J. S. Huxley, *Essays in Popular Science* (London: Penguin).

Huxley, J. S. (1938a) 'Darwin's Theory of Sexual Selection and the Data Subsumed by it in the Light of Recent Research', *American Naturalist*, vol. 72, pp. 416–33.

Huxley, J. S. (1938b) 'The Present Standing of the Theory of Sexual Selection', in Gavin de Beer (ed.), *Evolution* (Oxford: Clarendon Press).

Huxley, J. S. (1942) *Evolution: The Modern Synthesis* (London: George Allen & Unwin).

Huxley, J. S. (1959) Introduction, in P. Teilhard de Chardin, *The Phenomenon of Man* (London: Collins).

Huxley, J. S. (1964) *Essays of a Humanist* (London: Chatto & Windus).

Huxley, J. S. (1968) *The Courtship Habits of the Great Crested Grebe*, with a Foreword by Desmond Morris (London: Jonathan Cape).

Huxley, J. S. (1972) *Memories I* (Harmondsworth: Penguin).

Lovejoy, A. O. (1936) *The Great Chain of Being: A Study in the History of an Idea*; reprinted (1960) (New York: Harper).

Mayr, E. (1942) *Systematics and the Origin of Species from the Viewpoint of a Zoologist* (New York: Columbia University Press).

Mayr, E. (1980) 'Prologue: Some Thoughts on the History of the Evolutionary Synthesis', in E. Mayr and W. B. Provine (eds), *The Evolutionary Synthesis: Perspectives on the Unification of Biology* (Cambridge, Mass: Harvard University Press).

4 Scientific Work by Sir Julian Huxley, FRS

E. B. Ford, FRS

Sir Julian Huxley and I first met in 1921 when he became my Tutor at Oxford. We soon developed a joint piece of research to which he always attached much importance. Fundamentally, we showed that, in addition to the environment, genes can control the time of onset and rate of development of processes in the body of animals and plants (Ford and Huxley, 1927). Indeed, he and I were so much involved in that study that it is difficult after this length of time to apportion our respective contributions to it.

Huxley had seen a piece of breeding work carried out by Mrs Sexton at the Marine Biological Laboratory at Plymouth in which she showed that red eyes in the Amphipod Crustacean *Gammarus chevreuxi* segregate as a simple recessive to the normal black. He and I, realising the possibilities of such variation in the ommatidia, then demonstrated that the first colouring in the eye is due to a red pigment and that this is later concealed by the intensely black melanin. As a result of rather detailed study, we showed that this is controlled by genes governing variation in the inception and rate of development of structural and other features, affected also by temperature.

That concept was not reached until later, but one aspect of it was earlier approached by Goldschmidt in relation to his study of sex determination in the moth *Lymantria dispar*; an analysis which Sir Cyril Clarke has in the last few years shown to be incorrect (Clarke and Ford, 1980; 1982; 1983).

At the period of our work on 'rate-genes', Huxley was a Fellow of New College, so that the studies on *Gammarus* were principally carried out in the Department of Zoology at Oxford. However, we also examined the frequencies of the genotypes in a wild population. *Gammarus chevreuxi* is an exceedingly localised species, but it is, or was, abundant in the brackish water of Chelsdon Meadow, a salt-marsh near Plymouth. Consequently Huxley and I visited Plymouth several times and stayed there together. On one occasion we took

41

with us Dr Wolsky whom Huxley had met at the Biological Station at Lake Balaton.

It was of course obvious to us from the start that the existence of genes affecting rates of development must have an important impact upon evolution, allowing selection to favour either slow, or rapid, development in body size, individual structures and physiology. This therefore was an aspect of our finding in which we took particular interest, even in the earliest stages of our work. One feature of it, to which we drew special attention at the time, may briefly be mentioned here.

As generally recognised, human evolution has involved an unprecedented type of selection, one favouring intelligence. It has required a slowing down of development: a condition that makes it possible to lengthen the fetal and subsequent periods, so allowing extended opportunities for learning in childhood and for the delayed closing of the cranial sutures. This has necessitated the appropriate operation of 'rate genes'. Also there is the elimination of intra-uterine competition. That situation is achieved because mankind has, on the average, come to produce but one young at a birth (Ford and Huxley, 1927).

We are here led to two other features related to the evolving human line, for they particularly engaged Huxley's attention. One of these is the discovery, in which he was involved, that the dimorphism of the ability, or inability, to taste phenylthiourea has been maintained from a remote common ancestor, both in the human line and in that of the Great Apes. What the balanced genetic advantages may be we do not yet know, but of their reality there can be no doubt (Fisher *et al.*, 1939). It is certainly astonishing that such a polymorphism can be maintained for so long.

Secondly, I cannot avoid referring to an occurrence that Huxley would have wished me to mention: the possible light thrown on art-history when a baby gorilla, seeing its shadow thrown on a white wall by a powerful illumination, traced with its finger the demarcation between light and dark. It clearly expected some sensory distinction between the area illuminated and that in shadow. Huxley regretted that its finger was not dirty enough to leave a drawing (Huxley, 1942).

One thinks, in this matter, of sunlight in early morning, or evening, flooding into a rock shelter and casting on the wall the shadow of an inhabitant. This could perhaps add to the situations stimulating art.

Experimental embryology was one of Julian's chief interests in his earlier years (Huxley, 1932), carrying out much of the work himself.

Indeed, he invented an apparatus for keeping the two poles of a blastula at different temperatures. By this means he could accentuate distinctions between the sizes of the cells.

He later worked on relative growth. This was related to 'rate genes', and made it possible to analyse the varying relationships of one part of the body to another in a way that had not been successful previously. Thus, when he took y as the weight of an organ, say a limb or the liver, and x as that of the rest of the body, the changing relationship of y to x could be expressed as $y = bx^k$, where k is the measure of differential growth and b is a constant (Huxley, 1932). It then became apparent that certain species could be bimodal with respect to k: thus measuring a polymorphism, as in earwigs with large or small forceps.

Huxley's work, a little later, is perhaps particularly well known for his observations on courtship in birds, especially on the Louisiana heron, *Hydranassa tricolor*, and the Lesser egret, *Egretta thule*, while he made detailed studies on the habits of water-fowl, particularly at the lake in Blenheim Park (Huxley, 1923).

This, however, is only one aspect of his analysis of adaptation, depending on accurately carried-out field studies. One must refer also to his analysis of a wide range of experimentation and critical observation by others, demonstrating his remarkable knowledge of the relevant literature.

The term *cline* was introduced by Julian Huxley (Huxley, 1938; 1939) for a character gradient. It indicates the situation in which some animal or plant quality changes gradually and continuously over a large area. This is a useful expression, now widely employed, as it was by Huxley himself, both in giving an account of his own work and in analysing that of others. The features involved may be of the most diverse kinds, and generally depend upon adaptations to distinct environments. These can be such as are encountered in ascending a mountain, in passing from bog to dry country, or to subtle changes in habitat. Clines may comprise obvious features such as size, colour or proportion; or not obvious features, such as physiological qualities. They may indeed lead on to sub-speciation, reducing the viability of the heterogametic sex: the male in mammals, the female in birds.

Huxley found that Bergman's rule is usually applicable to warm-blooded animals: that is to say, body size generally increases with decreasing mean temperatures. He showed, indeed, both in animals and plants, that adaptive features are widely subject to clines. Thus wing-length in wrens, *Troglodytes*, increases by 1.0 per cent for every difference of 0.5°N in latitude. He accepted that clines may of course adjust to changing conditions, as in the homostyle primrose in the Sparkford neighbourhood of Somerset. There, thirty years ago, the homostyle plants formed a cline in passing from the status of a mutant to regions where they occupied 80 per cent, or so, of the plants. Today the maximum frequency is about 20 per cent.

Huxley was indeed much interested in plants with heterostyled and homostyled forms, favouring respectively outbreeding and inbreeding. Their adjustments to favourable, or unfavourable, conditions much attracted his attention. He was concerned to determine what comparable situations could have similar advantages in animals, and rightly picked upon the condition of fluctuation in numbers.

In this he followed with enthusiasm work on the butterfly *Militaea aurinia*. This is an annual species, and had been, superficially at any rate, very invariable for many years in a small population, then declining in numbers up to 1920. At that date, it evidently encountered an environment such as to favour it. For four years the numbers then increased to great abundance, involving also great variability. This situation suddenly changed to numerical stability accompanied by marked constancy in appearance. Huxley was much impressed that the new constant form differed from the former one: for selection is stricter in unfavourable conditions and relatively relaxed in favourable ones. We see here micro-evolution in action (Ford, 1975).

One noticed that Huxley seemed always able to produce relevant information on what might be under discussion. Concerning his thoughts on evolution, he considered a conclusion by J. B. S. Haldane that it is only in a society practising reproductive specialisation, so that most of the individuals are neuters, as in Social Hymenoptera, that very profound altruistic qualities, can be evolved. That is to say, of a type valuable to society but not to its possessors. Of this, Huxley remarks that unless we drastically alter our own type of reproduction, there is no hope of making the human species more decisively altruistic than it is at present.

There is no doubt that Julian Huxley's great work for evolution in general was accomplished by writing his monumental book *Evolution: the Modern Synthesis*.

In this he particularly stressed the significance of selection, when interest in that process had somewhat declined after Darwin's death. It was a synthesis to which he contributed to a remarkable extent by his own researches.

That book should be studied in its third edition (1974). The work, as originally published in 1942, remains intact, but was brought up to date in two ways. It includes a second edition, of 1963, by Sir Julian himself, with its own Introduction and Bibliography also intact. The third edition, however, contains a further additional Introduction and Bibliography by nine of Sir Julian's colleagues, bringing it again up to date (Huxley, 1974).

Finally I should like to return for a moment to our own early work on 'rate genes' in *Gammarus chevreuxi*. The time came when this needed to be written up in a joint paper, one that was published in 1927. On planning that account, Julian said, 'There is one point on which I insist. The paper must be by Ford and Huxley, not by Huxley and Ford'. No little gesture could have shown more clearly his consideration and his kindness.

References

Clarke, C. and Ford, E. B. (1980) 'Intersexuality in *Lymantria dispar* (L.) A Reassessment', *Proc. Roy. Soc. Lond.*, B206, pp. 381–94.

Clarke, C. and Ford, E. B. (1982) 'Intersexuality in *Lymantria dispar* (L.) A Further Reassessment', *Proc. Roy. Soc. Lond.*, B214, pp. 285–8.

Clarke, C. and Ford, E. B. (1983) '*Lymantria dispar* (L.) A Third Reassessment', *Proc. Roy. Soc. Lond.*, B218, pp. 365–70.

Fisher, R. A., Ford, E. B. and Huxley, J. S. (1939) 'Taste-testing the Anthropoid Apes', *Nature*, vol. 144, p. 750.

Ford, E. B. (1975) *Ecological Genetics*, 4th edn (London: Chapman & Hall).

Ford, E. B. and Huxley, J. S. (1927) 'Mendelian Genes and Rates of Development in *Gammarus chevreuxi*', *Brit J. exp. Biol.*, vol. 5, pp. 112–134.

Haldane, J. B. S. (1932) *The Causes of Evolution* (London: Longmans, Green).

Huxley, J. S. (1923) 'Courtship Activities in the Red-Throated Diver', *J. Linnean Soc. (Zoology)*, vol. 35, pp. 253–92.

Huxley, J. S. (1932) *Problems of Relative Growth* (London: Methuen).

Huxley, J. S. (1938) 'Clines: an Auxiliary Taxonomic Principle', *Nature*, vol. 142, pp. 219–20.

Huxley, J. S. (1939) 'Clines: an Auxiliary Method in Taxonomy', *Bijdragen tot de Dierkunde*, vol. 27, pp. 491–520.

Huxley, J. S. (1942) 'Origins of Human Graphic Art', *Nature*, vol. 149, pp. 637, 733.

Huxley, J. S. (1974) *Evolution: the Modern Synthesis*, 3rd edn (London: George Allen & Unwin).

5 Evolution Since Julian Huxley

Bryan C. Clarke, FRS

INTRODUCTION

There has been a lot of evolution since Julian Huxley, and I will have to be selective. I will take as my starting point the publication of that fascinating book, *Evolution as a Process*, edited by Huxley, Hardy and Ford in 1954, and concentrate on Huxley's main evolutionary interests, culled from his great work *Evolution: the Modern Synthesis*. They were:

(1) The extent of genetic variation;
(2) The power of natural selection;
(3) The origin of species;
(4) The mechanics of the evolutionary process.

Although these topics are interwoven, they can be considered one by one until, at the end of this essay, I will try for a 'Huxleian' synthesis.

THE EXTENT OF GENETIC VARIATION

It is not necessary to labour the point that the electrophoretic separation of proteins has revealed a great deal of genetic variation. Surveys of soluble proteins suggest that about 30 per cent of their coding loci are polymorphic (for reviews, see Powell, 1975; Nevo, 1978). Insoluble proteins, for example those contributing to physical structures much larger than themselves, seem to be rather less variable.

The discovery of ways to determine the sequences of nucleic acids has shown a situation that is even more remarkable. *Almost every gene varies from individual to individual* (see, for example, Nei, 1987). Genes coding for invariant proteins often differ in the third positions of their codons (when such nucleotide changes do not alter

46

the amino-acid sequence), or in their introns (pieces of DNA, within the coding region, that do not specify amino acids, and that are removed from an RNA message before protein synthesis) or in other non-coding pieces of DNA near the coding region. The data are yet too sparse to tell how many of these differences represent polymorphisms in the sense of Ford (1940), but it is clear that there is a phenomenal amount of genetic variation. Two individual chromosomes, taken at random from an outbreeding population, are likely to differ at one in every two or three hundred nucleotides (Nei, 1987). Since a human gamete contains about 3×10^9 nucleotides of nuclear DNA, two unrelated gametes taken at random are expected to differ in about ten million nucleotide sites. If each nucleotide variant is rare, two unrelated people are likely to differ in nearly twenty million sites. Each individual is indeed unique.

Clearly the amount of genetic variation within populations is very much larger than Huxley and his contemporaries supposed. This newly-discovered wealth of diversity has the interesting consequence that heterozygous advantage is eliminated as a general mechanism for the maintenance of variation. If virtually all individuals are heterozygotes there is little scope for selection against homozygotes (Lewontin *et al.*, 1978).

The field is open to alternative explanations of genetic diversity. Either we can suppose that most of the variation is selectively neutral, and exists because of a balance between mutation and random genetic drift (Kimura, 1983), or that it is maintained by some other form of balancing selection. At the moment, the best candidate among explanations alternative to drift is frequency-dependent selection.

We know that many selective agents can act in a frequency-dependent manner, causing the selective value of a phenotype to increase when it becomes rarer. Predators tend to over-eat common varieties of prey, and to neglect rare ones (Poulton, 1884; Allen and Clarke, 1968; Allen, 1988). Parasites probably become better adjusted to common varieties of host than to rare ones (Haldane, 1949; Clarke, 1979). Competitors can also act in a frequency-dependent manner (Christiansen, 1988). These forces work as efficiently on heterozygotes as on homozygotes.

I will come to the problem of deciding between evolutionary theories based on random genetic drift and those based on frequency-dependent selection when discussing the mechanics of the evolutionary process. Meanwhile it is worth mentioning that whereas Dobzhansky, in the 1941 edition of *Genetics and the Origin of Species*, regarded heterozygous advantage as the sole selective generator of polymorphism, Huxley, in *The Modern Synthesis* (1942), very clearly stated that frequency-dependent selection, which he called 'ecological balance', can be important in maintaining variation.

THE POWER OF NATURAL SELECTION

Theoreticians have tended to assume that the selective advantages and disadvantages of genetic variants, apart from those of lethals, are very small. Average selective coefficients of 0.001 or less are often postulated. Ecological geneticists in the tradition of Ford and Huxley, on the other hand, have been impressed by the precision with which organisms are adjusted to their environments, and have often found much larger differentials.

Notable studies in ecological genetics have been those of Kettlewell (1973) on melanic varieties of moths, Clarke and Sheppard (Clarke *et al.*, 1968) on mimetic butterflies, Cain and Sheppard (1954) on *Cepaea*, and Bradshaw and his colleagues (Bradshaw, 1971) on metal tolerance in plants. Huxley took a strong interest in the early stages of all these studies. They all gave evidence of strong selection (see Berry, 1977).

Recently Endler (1986) has made an encyclopaedic review of the literature and has tabulated selective coefficients measured in natural situations. Among undisturbed populations the median selective coefficient for polymorphic loci was about 0.3, and among disturbed, perturbed or stressed populations the median was about 0.5. Large values were also obtained for selection acting on quantitative characters. As Endler points out, the data may be strongly biased because ecological geneticists have concentrated on loci with large phenotypic effects, and because low and statistically insignificant selective coefficients are less likely to be reported. None the less it is clear that natural selective differentials for polymorphic loci can be large, and that theoretical treatments should take this possibility into account.

When we consider selection acting on a quantitative character, to which many loci may contribute, it is possible that even when selective differentials are strong the selection acting on individual loci may be weak. Obviously the strength of selection per locus depends upon the number of loci influencing the character, and upon the distribution of their effects. These are themselves matters of debate (see, for example, Thoday, 1977). Kimura (1983) believes that the average number of loci contributing to an average quantitative character is large, whereas Thoday (1977) believes it is relatively small. The experimental resolution of this disagreement will not come easily.

The problems are also acute at the level of variation in proteins and DNA, where the connections between changes in the molecules and changes in any phenotypes 'seen' by natural selection are generally obscure. If the variation is first detected at the molecular level, it is difficult to show that selective differences are due to the locus itself, rather than to closely linked loci. The difficulty can be overcome, but its solution is technically demanding (Clarke, 1975). None the less, strong selection acting on protein polymorphisms *has* been demonstrated experimentally (Clarke, 1975; Koehn *et al.*, 1983), and the human sickle-cell and glucose-6-phosphate dehydrogenase poly-. morphisms show that it can be a potent evolutionary force (Allison, 1964; Luzzato, 1979).

I know of no experiments on the population or ecological genetics of DNA variants that do not alter the amino-acids in proteins, apart from work on variants that change the numbers of genes. It is widely assumed, for example by Kimura (1983), that synonymous mutants in the third positions of codons, and variants in the non-coding regions of DNA, are selectively neutral. This assumption has not been tested, and it is worth noting that there are many ways by which such variants *could* be subject to selection.

Let us first consider synonymous mutants. The use of different codons for the same amino acid is not random (Grantham *et al.*, 1981). The frequencies of codons are related to the amounts of corresponding tRNAs and to the level of protein synthesis (Ikemura and Ozeki, 1982). These observations suggest that selection has acted on the choice of synonymous codons, and that the selection occurred because alterations in the third positions of codons affect the quantities of gene products. The availability of tRNAs may not be the only factor involved. Changes of nucleotides could alter the three-dimensional structure, and hence the stability, of the message. They could also affect the frequency of errors during translation by changing the stability of codon-anticodon binding.

Mutations in non-coding DNA can alter promoter sites, upstream controlling regions and enhancer sites, and thereby influence the rate of protein synthesis. Mutations in *any* part of the DNA may alter its three-dimensional structure, stability, availability to other molecules, speed of replication, susceptibility to restriction enzymes, susceptibility to methylation, and so on. All these influences are potentially subject to natural selection.

THE ORIGIN OF SPECIES

Huxley was of course greatly interested in the mechanisms by which species originate, and a large part of *The Modern Synthesis* is devoted to the problem. In general, the conclusions of 1941 still stand. Arguments for the importance of geographical (allopatric) speciation have been forcefully made by Mayr (1963), and arguments for the importance of sympatric and parapatric speciation have been made, with equal force, by White (1978). Most people now agree that all three can occur, but that allopatric speciation is the commonest mode.

There are two brilliant studies of speciation that deserve special notice. The first is by Carson and his colleagues on the picturewinged *Drosophila* of the Hawaiian islands (Carson and Kaneshiro, 1976, and review by Williamson, 1981). The Drosophilidae have undergone an extraordinary radiation in the archipelago, generating more than 500 species, possibly even 800. Carson and Kaneshiro have used species-specific rearrangements in the giant chromosomes of the salivary glands to work out the relationships between 100 species of picture-winged flies, and they have interpreted these relationships in terms of the geological history of the islands. The pattern is remarkable. There seem to have been orderly colonisations from north-west to south-east along the chain (i.e. from the older islands to the younger ones). Studies on the enzyme polymorphisms of the *planitibia* group produce a tree of resemblance roughly corresponding with that derived from the chromosomes, but with some discrepancies. Comparisons between species in morphology, chromosomes, and proteins clearly suggest that the different aspects of the organism evolve at different rates in different lineages. Some pairs of species are morphologically very similar, but chromosomally very different, some are chromosomally very similar but morphologically very different, some are chromosomally similar but differ in their enzymes, and in others the reverse is true. None of the characters seems to behave in a strictly clocklike manner. Similar discrepancies between morphological and enzymic divergence have been seen in the land snails (genus *Partula*) of the Society archipelago.

They also suggest an orderly series of colonisations from the older to the younger islands, and variations in evolutionary rates between lineages (Johnson *et al.*, 1986).

The second notable study is that by Grant and his colleagues on the finches of the Galapagos Islands (Grant, 1986). They have beautifully shown that many of the patterns of evolution and variation described by Lack (1947) can be explained by adaptation to particular ecological circumstances. The sizes and shapes of the birds' beaks are related to the sizes of available items of food, and the differences in beaks between species have been exploited in mating as aids to recognition. Grant provides strong evidence that competition between sympatric species causes divergences in beak sizes and diets. His work is probably the best-documented example of ecological, morphological and reproductive character-displacement.

Yang and Patton (1981) have studied protein polymorphisms in Darwin's Finches. There is here a reasonably good agreement between the tree of relationship based on proteins and that based on morphology.

THE MECHANICS OF THE EVOLUTIONARY PROCESS

If samples of a protein, say for example α haemoglobin, are taken from several species, and their amino-acid sequences are determined, it is possible to align the sequences, and to find what proportion of the amino-acids is shared by any pair of species. If every species is compared with every other, a matrix of proportions can be drawn up, and from this matrix a tree of resemblance can be constructed. It is a strange and beautiful fact that such matrices produce trees strongly resembling those based on comparative anatomy. This shows that protein molecules are not only adjusted to perform their functions, but also contain detailed information about the evolutionary relationships of the organisms that carry them. Stored within the product of a single gene is a potted history of its origins. The discovery of this phenomenon, by Zuckerkandl and Pauling (1965) has led to a whole new discipline, the study of molecular evolution.

Zuckerkandl and Pauling recognised two important and general patterns. First, the evolutionary rate, in terms of amino-acid substitutions per million years, is approximately constant for each protein. Rates measured in different parts of an evolutionary tree are roughly similar to each other. The rates for different proteins, however, diverge widely. The fastest-evolving proteins now known in euka-ryotes are the fibrinopeptides, and the slowest is Histone IV. Their rates differ by at least two orders of magnitude. When describing the rough constancy of rate within each class of protein, Zuckerkandl and Pauling coined the term 'the molecular clock'.

The second important observation was that substitutions between pairs of amino-acids differing greatly in their chemical properties ('radical substitutions') are much rarer than substitutions between chemically similar amino-acids ('conservative substitutions'). Zuck-erkandl and Pauling argued that radical substitutions are resisted by natural selection.

Since Zuckerkandl and Pauling's paper, it has become possible to study the nucleotide sequences of DNA and RNA as well as the amino-acid sequences of proteins. Nucleotide sequences, both coding and non-coding, also seem to behave in a clock-like way, and 'functionally-important' regions seem to evolve more slowly than 'functionally unimportant' ones. Presumably the functionally im-portant regions are more closely constrained by natural selection (for a review see Clarke *et al.*, 1986). The mitochondrial DNA of animals seems to be loosely constrained. It evolves unusually fast, and has proved to be extremely useful for determining relationships within genera (see Avise, 1986).

The apparent existence of molecular clocks has led several workers (e.g. Kimura, 1983; Nei, 1987) to argue that amino-acid and nucleo-tide substitutions are random events, that natural selection has a role only as a filter for removing disadvantageous mutations, and that most substitutions are selectively neutral. Proponents of this view contend that substitutions driven by natural selection should not occur at a constant rate, because they depend on sporadic changes in the environment.

Curiously, at the time when students of molecules were arguing that evolutionary rates are too constant to be attributable to natural selection, students of palaeontology were arguing that the rates are too *inconstant* to be so attributable. Eldredge and Gould (1972) put forward the theory of 'punctuated equilibria', pointing out that the fossil record shows short periods of rapid evolution, followed by longer periods of 'stasis'. For some reason Eldredge and Gould supposed that this pattern is incompatible with natural selection, despite the fact that variable rates of morphological evolution had been convincingly explained in Darwinian terms by Simpson (1944). In *The Science of Life* (Wells *et al.*, 1930) Huxley and his co-authors dealt robustly with an earlier saltational theory advanced by Hillaire Belloc. 'Contemporary biology', they said, 'has no need and no evidence for this grotesque hypothesis'. I am sure he would have felt the same about the arguments of Eldredge and Gould.

Even if such arguments against natural selection are weak, they do point to an interesting contrast. The progress of morphological evolution seems to be less regular than the progress of molecular evolution. The difference requires an explanation.

One has been suggested by Wilson (1976). He argued that the two kinds of characters are subject to different evolutionary forces. Amino-acid substitutions result from nucleotide substitutions in the coding regions. These, he argued, do not in general affect morphology, and are predominantly neutral to selection. Therefore they evolve at constant rates. Morphological changes, on the other hand, are more likely to be due to substitutions in restricted parts of the non-coding regions, parts that determine the quantities and dispositions of proteins rather than their sequences. These substitutions, he supposed, are subject to natural selection, and evolve at inconstant rates.

It transpires that Wilson's dual explanation may not be necessary. In the first place, molecular clocks are not as constant as they seemed at first. The earliest studies estimated rates over long periods of evolutionary time, and fluctuations over shorter periods were 'averaged out'. However, short-term fluctuations certainly occur (Langley and Fitch, 1974 and the work on flies and snails discussed above).

In the second place, molecular and morphological evolution are measured on different scales. A single nucleotide substitution among many would have little influence on a measure of molecular divergence, but it might have drastic effects on the phenotype.

In the third place, morphological characters are often influenced by many loci. The genes underlying these characters may, under some circumstances, be selected as much for their influence on the variance of a character as for their influence on its mean. Thus there will be no necessary correlation between the amount of evolutionary change at the level of the genes and the amount of change in the mean value of the phenotype. Molecular divergence due to natural selection can occur while the phenotypic mean remains constant. Of course the same is true of divergence due to random genetic drift.

While such arguments can help to explain discrepancies between molecular and morphological rates, they may not be enough by themselves to explain the roughly clock-like behaviour of molecules.

Can we find an explanation of 'clocks' if natural selection is the moving force of molecular evolution? I believe that we can, and that in so doing we may achieve a new synthesis between the theories of population and quantitative genetics.

These two disciplines have developed along paths that are more or less independent. Quantitative geneticists, faced with complex genetic systems, have been obliged to ignore the behaviours of individual loci. Population geneticists have concentrated on simple genetic systems because the algebra becomes impossible with more than two loci. Only now, by the use of powerful computers, are we able to study the individual behaviour of loci contributing to a quantitative character under selection. The computer models are producing some surprising results.

One such model (Clarke *et al.*, 1988) has shown the importance of a stochastic process that causes divergence when polygenic systems are subject to selection, and that can be clearly distinguished from random genetic drift. It is the random order in which mutations occur. Two isolated populations under the same selective pressures will diverge because different mutations will appear, and be selected, in each. The populations will diverge even if their other circumstances are identical. Under realistic conditions this selectively-driven stochastic process can be more important in causing divergence than random genetic drift.

Huxley fully appreciated the role of random mutation in bringing about divergence, a role which had been emphasised by Muller (1939; 1940) in a book edited by Huxley. Lack (1947) used mutation to explain differences between Darwin's Finches. Curiously, however, the selectively-driven process has been totally ignored by all students of molecular evolution (except Gillespie, 1984).

Our models (Clarke *et al.*, 1988) suggest that when selection acts on quantitative characters determined by several or many genes, the randomness of mutation can generate a rough molecular clock, driven entirely by the selection. Gillespie (1986) has reached similar conclusions. Models of this kind offer the promise of a selective theory of molecular evolution and, at the same time, of a bridge between population and quantitative genetics, and morphological and molecular evolution.

References

Allen, J. A. (1988) 'Frequency-dependent Selection by Predators', *Phil. Trans. Roy. Soc. Lond. B.*, vol. 319, pp. 485–503.

Allen, J. A. and Clarke, B. C. (1968) 'Evidence for Apostatic Selection by Wild Passerines', *Nature, Lond.*, vol. 220, pp. 501–2.

Allison, A. C. (1964) 'Polymorphism and Natural Selection in Human Populations', *Cold Spring Harbor Symp. Quant. Biol.*, vol. 29, pp. 137–49.

Avise, J. C. (1986) 'Mitochondrial DNA and the Evolutionary Genetics of Higher Animals', *Phil. Trans. Roy. Soc. Lond. B.*, vol. 312, pp. 325–42.

Berry, R. J. (1977) *Inheritance and Natural History* (London: Collins).

Bradshaw, A. D. (1971) 'Plant Evolution in Extreme Environments', in F. R. Creed (ed.), *Ecological Genetics and Evolution* (Oxford: Blackwell).

Cain, A. J. and Sheppard, P. M. (1954) 'Natural Selection in *Cepaea*', *Genetics*, vol. 39, pp. 89–116.

Carson, H. L. and Kaneshiro, K. Y. (1976) '*Drosophila* of Hawaii: Systematics and Ecological Genetics', *Ann. Rev. Ecol. Syst.*, vol. 7, pp. 311–45.

Christiansen, F. B. (1988) 'Frequency Dependence and Competition', *Phil. Trans. Roy. Soc. Lond. B.*, vol. 319, pp. 587–600.

Clarke, B. (1975) 'The Contribution of Ecological Genetics to Evolutionary Theory: Detecting the Effects of Natural Selection on Particular Polymorphic Loci', *Genetics*, vol. 79, pp. 101–3.

Clarke, B. (1979) 'The evolution of Genetic Diversity', *Proc. Roy. Soc. Lond. B.*, vol. 205, pp. 453–74.

Clarke, B., Robertson, A. and Jeffreys, A. J. (eds) (1986) *The Evolution of DNA Sequences* (London: The Royal Society).

Clarke, B., Shelton, P. R. and Mani, G. S. (1988) 'Frequency-dependent Selection, Metrical Characters and Molecular Evolution', *Phil. Trans. Roy. Soc. Lond. B.*, vol. 319, pp. 631–40.

Clarke, C. A., Sheppard, P. M. and Thornton, I. W. B. (1968) 'The Genetics of the Mimetic Butterfly *Papilio memnon*', *Phil. Trans. Roy. Soc. Lond. B.*, vol. 254, pp. 37–89.

Dobzhansky, T. (1941) *Genetics and the Origin of Species* (New York: Columbia University Press).

Eldredge, N. and Gould, S. J. (1972) 'Punctuated Equilibria: an Alternative to Phyletic Gradualism', in T. J. M. Schopf (ed.), *Models in Paleobiology* (San Francisco: Freeman).

Endler, J. A. (1986) *Natural Selection in the Wild* (Princeton: University Press).

Ford, E. B. (1940) 'Polymorphism and Taxonomy', in J. Huxley (ed.), *The New Systematics* (Oxford: University Press).

Gillespie, J. H. (1984) 'Molecular Evolution over the Mutational Landscape', *Evolution*, vol. 38, pp. 1116–129.

Gillespie, J. H. (1986) 'Natural Selection and the Molecular Clock', *Mol. Biol. Evol.*, vol. 38, pp. 138–55.

Grantham, R., Gautier, C., Gouy, M., Jacobzone, M. and Mercier, R. (1981) 'Codon Catalog Usage is a Genome Strategy Modulated for Gene Expressivity', *Nucl. Acids Res.*, vol. 9, pp. r43–r74.

Grant, P. R. (1986) *Ecology and Evolution of Darwin's Finches* (Princeton: University Press).

Haldane, J. B. S. (1949) 'Disease and Evolution', *Ricerca scient. Suppl.*, vol. 19, pp. 68–76.

Huxley, J. S. (1942) *Evolution: the Modern Synthesis* (London: George Allen & Unwin).

Huxley, J., Hardy, A. C. and Ford, E. B. (eds) (1954) *Evolution as a Process* (London: George Allen & Unwin).

Ikemura, T. and Ozeki, H. (1982) 'Codon Usage and Transfer RNA Contents: Organism-specific Codon-choice Patterns in Reference to the Isoacceptor Contents', *Cold Spring Harbor Symp. Quant. Biol.*, vol. 47, pp. 1087–96.

Johnson, M. S., Murray, J. and Clarke, B. (1986) 'An Electrophoretic Analysis of Phylogeny and Evolutionary Rates in the Genus *Partula* from the Society Islands', *Proc. Roy. Soc. Lond. B.*, vol. 227, pp. 161–77.

Kettlewell, H. B. D. (1973) *The Evolution of Melanism* (Oxford: Clarendon Press).

Kimura, M. (1983) *The Neutral Theory of Molecular Evolution* (Cambridge: University Press).

Koehn, R. K., Zera, A. J. and Hall, J. G. (1983) 'Enzyme Polymorphism and Natural Selection', in M. Nei and R. K. Koehn (eds), *Evolution of Genes and Proteins* (Sunderland, Mass.: Sinauer).

Lack, D. (1947) *Darwin's Finches* (Cambridge: University Press).

Langley, C. H. and Fitch, W. M. (1974) 'An Examination of the Constancy of the Rate of Molecular Evolution', *J. Mol. Evol.*, vol. 3, pp. 161–77.

Lewontin, R. C., Ginzburg, L. R. and Tuljapurkar, S. D. (1978) 'Heterosis as an Explanation for Large Amounts of Genetic Polymorphism', *Genetics*, vol. 88, pp. 149–70.

Luzzato, L. (1979) 'Review: genetics of Red Cells and Susceptibility to Malaria', *Blood*, vol. 54, pp. 961–76.

Mayr, E. (1963) *Animal Species and Evolution* (Cambridge, Mass.: Harvard University Press).

Muller, H. J. (1939) 'Reversibility in Evolution Considered from the Standpoint of Genetics', *Biol. Rev.*, vol. 14, pp. 261–80.

Muller, H. J. (1940) 'Bearing of the *Drosophila* Work on Systematics', in J. Huxley (ed.), *The New Systematics* (Oxford: University Press).

Nei, M. (1987) *Molecular Evolutionary Genetics* (New York: Columbia University Press).

Nevo, E. (1978) 'Genetic Variation in Natural Populations: Patterns and Theory', *Theor. Popul. Biol.*, vol. 13, pp. 121–77.

Poulton, E. B. (1884) 'Notes Upon, and Suggested by, the Colours, Markings and Protective Attitudes of Certain Lepidopterous Larvae and Pupae, and of a Phytophagous Hymenopterous Larva', *Trans. Ent. Soc. Lond.*, pp. 27–60.

Powell, J. R. (1975) 'Protein Variation in Natural Populations of Animals', *Evolut. Biol.*, vol. 8, pp. 79–119.

Simpson, G. G. (1944) *Tempo and Mode in Evolution* (New York: Columbia University Press).

Thoday, J. M. (1977) 'Effects of Specific Genes', in E. Pollak, O. Kempthorne and T. B. Bailey (eds), *Proc. Int. Conf. Quantitative Genetics* (Ames: Iowa State University Press).

Wells, H. G., Huxley, J. S. and Wells, G. P. (1930) *The Science of Life* (London: Cassell).

White, M. J. D. (1978) *Modes of Speciation* (San Francisco: W. H. Freeman).

Williamson, M. (1981) *Island Populations* (Oxford: University Press).
Wilson, A. C. (1976) 'Gene Regulation in Evolution', in F. Ayala (ed.),
   Molecular Evolution (Sunderland, Mass.: Sinauer).
Yang, S. H. and Patton, J. L. (1981) 'Genic Variability and Differentiation in
   Galapagos Finches', *Auk*, vol. 98, pp. 230–42.
Zuckerkandl, E. and Pauling, L. (1965) 'Evolutionary Divergence and
   Convergence of Proteins', in V. Bryson and H. J. Vogel (eds), *Evolving
   Genes and Proteins* (New York: Academic Press).

6 Julian Huxley and the Rise of Modern Ethology

R. I. M. Dunbar

During the first three decades of this century, Julian Huxley carried out a series of field studies on the courtship behaviour of water birds. These studies have not only stood the test of time, but they are in many ways models of ethological field work. In fact, the debt that ethology owes to Huxley has not often been recognised, mainly I suspect because Huxley's active research in this area occurred so much earlier than the main flowering of ethology after the Second World War. Yet even a casual reading of Huxley's early papers reveals a surprising number of respects in which he anticipated the conceptual developments that were to occur later in the work of the classical ethologists.

In this chapter, I shall try to show just how great this debt to Huxley in fact is. I shall begin with some examples of Huxley's own work and the contributions that he made to the theory and practice of ethology. I shall then go on to consider the kinds of field work that we are doing now in order to illustrate the extent to which we are still in many ways following in Julian Huxley's intellectual footsteps. In this respect, I do not want to suggest that what we are doing now is merely an imitation or elaboration of the work that Huxley himself did more than half a century ago. But I do want to suggest that contemporary interests not only owe their roots to Huxley's work, but also that they are developments – in some respects, even major revolutions of thinking – that Julian Huxley would very much have approved of.

HUXLEY'S CONTRIBUTION TO CLASSICAL ETHOLOGY

The studies of the courtship behaviour of grebes, divers and other waterbirds which Huxley carried out during the early years of this century (see Huxley, 1914; 1923) are impressive both for their meticulous attention to detail and for the acuteness of their understanding of the birds' natural histories. But, as an ethologist, I am struck even more forcefully by the surprising number of concepts that Huxley discusses in these papers which were to feature prominently in the theoretical vocabulary of classical ethology four decades later.

It was Huxley, for example, who first described and named the process of 'ritualisation' whereby a behaviour pattern or morphological feature becomes adapted for use in a context to which it is not directly relevant. This process invariably involves an elaboration and exaggeration of the feature concerned, the classic examples being the exaggerated preening and head-shaking seen in the courtship rituals of the grebe, the head dipping of mute swans and the presentation of nesting material or food that features in the courtship of many species of birds (Huxley, 1923). From his detailed personal knowledge of these species' behaviour, Huxley realised that most of these curious behaviour patterns are simply elaborate versions of behaviour that occurs quite naturally in other contexts. Head-shaking, for example, is simply a stylised form of the behaviour that grebes show on emerging from the water after a dive. In this context, head-shaking obviously functions to remove water from the face and head feathers, and the behaviour pattern has been incorporated into the courtship sequence even though the head has not been submerged. What characterises ritualisation, then, is the use of a behaviour pattern in an apparently irrelevant context, and particularly one in which emotional tension and motivational conflict are at a high pitch.

Huxley's emphasis on the motivational conflicts in such contexts provides another example of his contribution to the theoretical structure of ethology, for his 1923 paper provides the first unequivocal descriptions of what were later to be called 'displacement activities'. Indeed, he specifically suggests that such apparently irrelevant behaviours are the result of the 'sparking over' of powerful emotions when these are unable to find expression through normal motor channels at times of motivational conflict. The 'sparking over' theory remained one of the three main explanations for displacement activities well into the post-war period, though it is usually associated with the name of another of the leading classical ethologists, Niko Tinbergen (see Tinbergen, 1952). Although the 'sparking over' theory of displacement activities has long since been refuted (see Hinde, 1966), it is still considered to be important enough in an historical sense to merit mention even in contemporary textbooks (e.g. Slater, 1985).

One last example concerns the nature of territoriality. From a series of observations on the way in which coots adjusted their territories in response to changes in the density of birds on a pond, Huxley (1934) concluded that territories are like elastic discs. When the density of animals is low, competition for territories is also low and pairs can spread themselves out in large ranges; as the density rises, however, pressure from newcomers forces residents into ever smaller areas and territories become compressed. Although there has continued to be some debate over Huxley's concept, the notion that territories respond like elastic discs to the density of intruders has been widely confirmed in field studies of species ranging from dunlin (Holmes, 1970) and sanderling (Myers *et al.*, 1979) to colobus monkeys (Dunbar, 1987).

Huxley's perceptiveness in identifying so many of the concepts that were to lie at the heart of classical ethology can, I think, be attributed to his emphasis on the careful observation of the animal in its natural habitat. In his paper on courtship rituals, he observed that:

> Those who have opportunity, patience and a good glass, and are willing to take full notes, will find that steady observation of almost any species of bird at the beginning of the breeding season, particularly if a single pair can be followed throughout, will bring results which may be of very considerable interest, both from the standpoint of pure biology and also from that of comparative psychology, as well as being in itself a very fascinating occupation. (Huxley, 1923, p. 266)

It was this emphasis on naturalistic observation that was to become the hallmark of ethology during the 1930s: careful painstaking observation of natural behaviour was to be the one feature that ethologists from Konrad Lorenz onwards were to emphasise time and again. It was their appreciation of the natural world for its own sake and their enthusiasm for watching animals in the wild that clearly differentiated them from their laboratory-based contemporaries in comparative psychology. I do not wish to suggest that the later ethologists borrowed their approach from Huxley: indeed, Huxley was by no means the first to extol the delights and virtues of natural history, for there had been a long tradition of skilled amateur naturalists in Britain from the time of Gilbert White's studies at Selborne (White, 1789) to the meticulous observations of the naturalist Edmund Selous (Selous, 1901). None the less, Huxley did introduce to this essentially descriptive natural history an important new dimension based firmly on a detailed understanding of Darwin's theory of evolution by natural selection. In the best traditions of explanatory (as opposed to purely descriptive) science, he sought to use Darwin's theory to understand the behaviour of animals and to use the behaviour of the animals to develop new insights into evolutionary theory.

This tendency to develop an interactive approach between observation and theory led him to emphasise a second feature that the later ethologists were to set great store by, namely the functional analysis of behaviour. The idea that we could study behaviour from an evolutionary point of view in exactly the same way that comparative anatomists studied morphology was far from new: it had been voiced by Darwin himself (e.g. Darwin, 1872) as well as many others among his contemporaries. But in the decades after Darwin's death, functional analyses had become increasingly speculative as armchair theorists tried to explain everything in terms of adaptation, often with scant regard for the natural history and biology of the animals concerned. Huxley (1942) himself cites as one of the worst cases of this the example of Thayer (1909) who argued that the pink colour of flamingoes was designed to camouflage them against the setting sun as they flew home to roost in the evening. Huxley sought to reintroduce a more rigorous approach to functional analyses, an approach that was neatly encapsulated in his catch-phrase 'Structure first, function afterwards' (Huxley, 1914, p. 492).

Huxley insisted that if we are to make any sense of the behaviour of animals, it is critical that we understand exactly how the animal uses that behaviour in its daily life; then, once we have arrived at a clear hypothesis, we should set about testing this suggestion with all the rigour at our disposal, trying both to exclude alternative null hypotheses and to show that such behaviour occurs only in those contexts (or species) where its performance is crucial in allowing animals to solve the presumed problem in order to survive and/or reproduce more effectively.

Perhaps in the process of thinking about the problems inherent in evolutionary explanations of behaviour, Huxley (1942) was later led to differentiate what he termed the three different senses that can be attributed to any biological 'fact'. In considering a behavioural or morphological feature, he pointed out, we can offer explanations or descriptions at three very different levels. We can give a mechanical (or physiological) explanation describing how the particular phenomenon works; we can give a functional or adaptive explanation by identifying the problem that such a feature enables the animal to solve; and, finally, we can offer an evolutionary or historical explanation of the sequence of changes in, for example, the behavioural repertoires of the species' ancestors that produced the phenomenon we see now. This stress on the different facets of biological phenomena was later to taken up by Tinbergen (1963) in his authoritative statement on the methods of ethology. To Huxley's three dimensions, Tinbergen added only a fourth, the study of development or ontogeny.

One last theoretical concept that Huxley introduced is worth mentioning because it had a profound impact on later developments that led up to some of the major interests that dominate our thinking today. In a brief note published in 1959, Huxley (1959) used the term 'grade' to refer to the characteristic lifestyle of a set of animals (or species) that occupy a particular ecological niche (although I use the term niche here in a rather broad sense). This idea was subsequently taken up and developed by John Crook in a seminal series of papers on the evolution of social systems in weaver birds (Crook, 1962; 1965) and primates (Crook and Gartlan, 1966; Crook, 1970). He argued that adaptation to the requirements of a particular niche can be expected to have predictable consequences for patterns of dispersion which, in turn, would be likely to favour particular mating and grouping patterns. The suggestion that ecology might be the driving force behind much of social evolution generated a great deal of interest and stimulated two decades of field work. Indeed, the basic idea is still highly influential and, implicitly or explicitly, lies behind most of the research currently being done on the behaviour of freeliving animals.

I would not want to appear to be suggesting that these later developments owe their origins wholly to Huxley's earlier work. Huxley himself later recognised that his original concepts were greatly refined by ethologists like Lorenz and Tinbergen during the 1930s and 1940s (see Huxley, 1966). Indeed, he goes so far as to admit that in some cases he had himself failed to appreciate the full significance of some of the phenomena that he had described and that some of these might have been buried in the dusty obscurity of learned journals for all time had they not been taken up and elaborated by Lorenz and his colleagues – ritualisation being a particular case in point.

SOCIAL BIOLOGY TODAY

The intrinsic difficulty of studying the functional and evolutionary (i.e. historical) aspects of behaviour resulted in a general dissatisfaction with such analyses during the 1960s and a consequent tendency for a greater emphasis to be placed on studies of motivation, proximate mechanisms and development. Only in the area that became known as socio-ecology was a functional perspective in any sense paramount, but this remained a moderately small component of the research effort in ethology as a whole. Matters changed dramatically during the later 1970s, however, as a consequence of the sociobiological revolution. By showing how the mathematics of population genetics could be applied to behaviour, sociobiology made it possible to introduce a more rigorous (and therefore more testable) approach to the study of function. At the root of this lay the crucial realisation that the criterion by which we should judge evolutionary success is the number of copies of a given gene that an individual contributes to future generations. This, in essence, is the notion of the 'selfish gene' which Richard Dawkins introduced in his book of the same title (Dawkins, 1976). It reminds us that whenever we are considering the evolutionary consequences of behaviour, we must always ask ourselves what would happen to a specific gene were it the case that such behaviour was determined ontogenetically by a single gene?

Unfortunately, this way of expressing the fundamental principles of Darwinian evolutionary theory is open to misinterpretation, since it seems to imply that behaviour is genetically determined in a one-gene/one-character sense (or at the very least that the only kinds of behaviour that are interesting are those which are so determined). But the way in which the notion of a selfish gene is used is essentially metaphorical: it is a heuristic device intended to make us focus our attention on the fact that one thing, and one thing only, makes evolution possible, namely the genes that are passed on from one generation to the next.

While it is certainly true that some sociobiologists seem to have a one-gene-for-one-behaviour view of the world, the insistence that sociobiologists in general are obliged to think in this way strikes me as rather odd. For one thing, it severely limits the range of biological phenomena that will be of interest to sociobiology since we know that a great many of the day-to-day decisions made not only by human beings but also by many higher animals are flexible, learned re-

sponses; yet, it remains the case that all of these behaviours necessarily influence genetic fitness since they must affect the organism's
ability to survive and reproduce. A second problem with the genetical view of sociobiology is that in all honesty we simply do not know
what role genes play in the development of specific behaviour patterns.

But does this really matter? I would argue not, for what is important from a sociobiological point of view is the functional consequences that all behaviour patterns necessarily have, not their
ontogeny. As Huxley stressed all those years ago, we cannot draw
inferences from function about other facets of a biological phenomenon. So long as there is some mechanism for closing the evolutionary
'loop' – i.e. for ensuring that those behavioural strategies which allow
organisms to produce many descendants are the ones that get passed
on to the next generation – then selection will work equally effectively whether or not the behaviour in question is directly underwrit-
ten by genes. Obviously, the great advantage of assuming a pure
form of genetic determinism is that the loop is closed directly by the
gene itself: it produces the behaviour pattern that allows a direct copy
of itself to be passed on to the next generation, where it can in its turn
produce a new copy of the same behaviour pattern. But we can
achieve the same effect in many other ways, and this, as I understand
it, is the whole point of Dawkin's (1976) discussion of 'memes': his
point was that we should not be constrained to thinking in terms of
the conventional biological replicators that we happen to have encountered here on the planet Earth. Selection and adaptation, as
universal phenomena, will occur with any substrate that is capable of
replicating itself with reasonable fidelity (Dawkins, 1984).

At a cognitively advanced level, of course, behavioural replication
in this sense can occur through learning. Where learning abilities are
sophisticated, mere copying of other individuals' behaviour may not
be necessary, for an intellectually sophisticated organism may be able
to solve problems on its own initiative from first principles by trial
and error. All that is necessary for fidelity of transmission to occur
between generations is a defined goal-state for the organism to aim at
and a brain of sufficient complexity to allow the organism both to
detect departures from the optimal state and to discover how to
restore its internal equilibrium by adjusting its behaviour. The kinds
of rules required can be very simple: 'Repeat what you did last time if
it worked; try something different if it did not'. Likewise, the inbuilt
goal-states need not be particularly complex: I suspect that we could

probably get away with the handful of conventional motivations that psychologists once referred to as 'drives' (i.e. hunger, thirst, comfort, sex, etc.). So long as these two components can be genetically programmed, then all of an animal's behaviour can be entirely flexible yet geared to and driven by conventional considerations of evolutionary (i.e. genetic) fitness.

Adopting this view has had two important advantages. First, sociobiology has allowed us to unify all aspects of an organism's biology within a single framework. This in principle allows us to compare directly the evolutionary advantages and disadvantages of options that an organism might be faced with, even though these are in completely different dimensions of its biology. Thus, we could now evaluate the relative advantages of, say, producing more embryos or investing more heavily in parental care. Moreover, the fact that we can make use of the formal theories of population genetics greatly increases the precision (and hence the power) of hypothesis testing. Secondly, in doing so, we can rely on the distinction between function and ontogeny to avoid having to concern ourselves with problems about the genetic determination of behaviour that are at present insoluble. This frees us from the geneticist's dilemma of having to establish the precise nature of the genetic input before embarking on any functional analyses.

This is not, of course, to say that questions about either the processes of development or the genetic control of behaviour are unimportant or uninteresting. It is simply to say that, contrary to the expressed views of some critics of sociobiology, we do not have to worry about them in order to ask meaningful questions about function.

CONTEMPORARY FIELD ETHOLOGY

Contemporary studies of behaviour differ in three important respects from those carried out both by Huxley himself and by the classical ethologists. The first is a deliberate shift away from the study of behaviour itself to the study of the relationships that the animals use the behaviour to mediate. This step back up to a higher dimension makes questions about function inevitable, since the relationships provide the functional context for behaviour patterns: an analysis of the nature of a relationship is by its very nature an analysis of the functional end of the behaviour patterns involved. The second respect in which contemporary studies differ from those of Huxley and the classical ethologists lies in their emphasis on quantitative data: counting rather than merely describing is the modern ethologist's main concern. Although the enumeration of results was introduced by Tinbergen (1963) at quite an early stage in the history of ethology, this was mainly in the context of field experiments rather than purely observational data. It is the use made of quantitative data for detailed statistical analysis that differentiates contemporary ethology from its predecessors. The third feature is closely related to this and lies in the use of powerful theories to generate detailed hypotheses about the behaviour of animals that can then be tested in a very rigorous way against data from the real world. I shall try to illustrate these features of contemporary studies by concentrating on just two examples. I hope that doing so will give a more vivid impression of the scope and depth of detail that field studies now aspire to.

One of the characteristic features of contemporary studies is the way they use theory as a tool for focusing attention on interesting questions. It has been noted, for example, that in many species whose males and females mate monogamously and cooperate in rearing the offspring, older offspring often remain with the parents to help rear the next litter (e.g. Tasmanian native hen: Ridpath, 1972; Florida scrub jay: Woolfenden, 1975; acorn woodpecker: Stacey, 1979; jackals: Moehlman, 1979; tamarin monkeys: McGrew and McLuckie, 1986). Such behaviour is surprising from a Darwinian point of view because, all other things being equal, an individual contributes most genes to the next generation by breeding itself. It is, of course, the case that any individual who helps to rear its siblings necessarily contributes to the next generation that proportion of its genes that it shares in common with its siblings (a rather loose way of phrasing Hamilton's [1964] notion of inclusive fitness); but since the likelihood of a given gene being propagated via this route depends on the probability that they actually do share the gene in question by descent from the same ancestor, the gains via what is usually known as kin selection have to be considerable to outweigh the fact that this probability is, for most vertebrates, at best 0.5. In other words, for every offspring that an animal forgoes by not breeding in order to help its parents, the parents must produce at least two extra offspring as a direct result of the animal's help if it is to benefit by kin selection alone. (For a lucid account of inclusive fitness and the theory of kin selection, see Graffen, 1984.) Clearly, the circumstances under which it will be worth an animal's while giving up personal reproduction are likely to be rather limited. The most likely context is where suitable breeding territories are in short supply and/or food supplies are so patchy that the animal risks not being able to breed at all. Such conditions are likely to be especially common in rather impoverished habitats or those whose quality fluctuates dramatically from year to year.

Steven Emlen has tested this prediction on bee-eaters, a group of insectivorous birds that breed in large colonies in the sandy banks of rivers in African savannah habitats (see Emlen, 1982). He found that the frequency of helping behaviour by unpaired birds was inversely related both to the rainfall in that year and to an index of insect abundance in the month before the onset of the breeding season. In other words, the availability of the food resources that a pair requires to rear its offspring determined whether the previous year's offspring attempted to breed for themselves or preferred to cut their losses, wait another year in the hope that conditions would improve and, in the meantime, gain some genetic fitness by helping their parents rear another brood. Emlen (1982) also noted that the mortality of nestlings is exceptionally high in the relatively impoverished and unpredictable foraging conditions provided by the savannah habitats in which this species lives. Hence, it is hardly surprising to find that breeding pairs which have helpers fledge significantly more of their chicks than pairs that do not. From the parents' point of view, the presence of unmated helpers is a considerable advantage.

However, the presence of so many unmated individuals creates a new kind of problem, especially for a colonially nesting species like the bee-eater. For individuals can expect to gain fitness at the expense of fellow colony-members if they can persuade other birds to rear additional offspring for them instead of their own. Emlen and Wrege (1986) noted that such additional gains can be achieved in either of two ways: (1) by females laying eggs in the nests of other individuals who are then left to rear them (a cuckoo-like strategy) and (2) by males mating with other males' mates so as to fertilise these females' eggs.

Emlen and Wrege (1986) were able to demonstrate that nest parasitism was indeed a common feature of white-fronted bee-eater behaviour: on average, 16 per cent of nests were parasitised, with about 7 per cent of the eggs in any given nest being laid by a female other than the owner. Significantly, females were very selective when laying their eggs in other individuals' nests. These were not chosen at random; rather, target nests were chosen to maximise the chances that the additional eggs would not be detected so as to give them the highest chances of being hatched successfully by the foster parents. Thus, 60 per cent of parasitic eggs were laid during the five-day period that the host female was actually laying her own eggs. Eggs laid during this narrow window had a significantly higher chance of hatching successfully than eggs laid just prior to the nest-owner's egg-laying period or during the subsequent 30 days of incubation. Eggs laid too early were tossed out of the nest by the owner when she began to lay her own eggs, while late-laid eggs, though not normally removed by the host, invariably failed to hatch because the foster parents could not continue to sit on them once their own nestlings had hatched and needed feeding.

An analysis of precisely which individuals laid eggs in other females' nests revealed two main categories of parasitic females. By far the largest group were females whose own breeding cycles had been disrupted (either because their mate had died or because the mate had failed to provide sufficient food for the female while she was confined to the nest in the pre-laying period). Emlen and Wrege (1986) were able to confirm this by experimentally disrupting the breeding cycles of a small number of females by blocking off the entrances to their nests once egg-laying had started: all these females subsequently laid parasitic eggs. The second group of parasitic females consisted of unpaired daughters who remained as helpers at their parents' nest: these sometimes laid eggs in the mother's nest, though in this case they also contributed to the rearing process.

Female bee-eaters exhibit a number of behavioural strategies that are clearly designed to minimise the risk of being parasitised. One of these is to remain in the nest burrow throughout the egg-laying period. Most female birds avoid the nest assiduously during the intervals between laying successive eggs and it is generally assumed that this helps to minimise the risk that predators will find the nest through being attracted to it by the female's movements in and around the nest. However, Emlen and Wrege (1986) suggest that, in the bee-eater's case at least, the risk of a nest being parasitised during the female's absence far exceeds any risk they run from predators. The second strategy that females use to avoid being parasitised is, of course, to remove any eggs from the nest that are not their own. It seems, however, that females cannot easily discriminate their own eggs from those of other females on the basis of visual cues alone since they only try to remove eggs laid before they themselves begin to lay. Emlen and Wrege (1986) were able to demonstrate this experimentally by adding artificial eggs to females' nests at various stages from seven days before egg-laying started to eleven days after. While 93 per cent of the 'eggs' put in the nest before the female started to lay were tossed out, none of those introduced after laying started were removed.

The concentration of breeding females at the colony site during the egg-laying period has significant implications for the males. In effect, they constitute a predictable and temporarily super-abundant resource. Since any extra eggs that a male can fertilise other than his own mate's are an added contribution to his fitness, Darwinian theory predicts that males will make every effort to mate with other females, providing that, in doing so, they do not seriously affect their mates' abilities to rear their own eggs. In addition, the concentration of receptive females is likely to attract unpaired males anxious to exploit any opportunity to make at least some contribution to the species' gene pool. Emlen and Wrege (1986) found that females leaving and entering their nest chambers unaccompanied were ten times more likely to be harassed by males than when they did so in the company of their mates. Although the frequency of successful copulation by these harassing males was quite low (estimated to account for only 0.7 per cent of the copulations that the average female engages in during a breeding season), none the less the amount of harassment suffered by a lone female was considerable. No less than 70 per cent of females leaving their nest chambers alone were harassed by males. Such harassment alone might prove to be sufficiently stressful to disrupt the female's breeding physiology.

Emlen and Wrege (1986) point to the concentration of females at colony sites as the proximate cause of this harassment and argue that the risk of a mate being fertilised by another male has been the key factor promoting monogamous pairbonding in this species. Her mate's continuous proximity has then allowed the female to remain in the nest chamber during the period immediately prior to egg-laying in order to prevent her nest being parasitised by other females since she is now able to count on the male to bring food to her throughout this period. At this point, we can begin to see a sequence of behavioural strategies emerging, each step of which is an attempt to solve a new problem created by a previous step in the chain. Thus, the birds are forced to concentrate on the few sites that provide them with suitable substrates in which to construct nests, but doing so leaves the females open to harassment by males; this problem can be resolved by the male staying with his mate and protecting her; and this in turn allows the female the opportunity to solve the problem of nest parasitism (itself a consequence of the opportunity provided by colonial nesting and the unpredictability of the environment).

This emphasis on the fact that an organism is constrained within a complex network of interacting biological factors that often conflict with each other is rapidly becoming one of the characteristic features of contemporary field work. It suggests that we cannot always usefully explore elements of the system in isolation, since many of the factors that give rise to and constrain an animal's behaviour in one respect are often a direct consequence of its attempts to solve other biological problems (see, for example, Dunbar, 1988b). Faced with such complexity, ethologists have over the past decade or so resorted increasingly to the use of modelling as a means of exploring the systemic nature of an animal's behavioural strategies. I should like to illustrate just one aspect of this use of modelling with some of my own work on the reproductive strategies of gelada baboons.

The most effective way of contributing genes to the species' gene pool is, of course, to breed and for most male mammals this means mating with as many females as is possible within the constraints imposed by the species' reproductive biology and the way in which the females distribute themselves around the habitat. In a species like the gelada, the females form small groups of up to a dozen individuals, together with their dependent young (Crook, 1966; Dunbar, 1984). A male's ability to breed thus depends on whether he can acquire and maintain control over a group of females. In the simplest of all possible worlds, the easiest way of doing this is by brute force: powerful males are able to prevent competitors from gaining access to the females, so that less powerful males are excluded. In the real world, however, the intensity of competition generated by a strategy based on outright aggression in a situation where a small proportion of the males can monopolise all the available resources (in this case, breeding females) creates conditions in which the costs of success spiral rapidly as the competition becomes more intense.

In the case of the gelada, males face two key problems in this respect (for further details, see Dunbar, 1984). The first is that the larger the group of females he can monopolise, the more males will be excluded from breeding and therefore the more intense the pressure that he will come under from males trying to evict him. This naturally means more fighting and greater risk of injury as well as greater risk of defeat and perhaps even premature death. The second problem is that the females themselves suffer increasingly from reproductive suppression as the size of the unit increases. This is mainly due to the fact that harassment and stress increases with the number of animals in the group and that high levels of stress disrupt the female's reproductive physiology (see Bowman *et al.*, 1978). The important consequence of this from the male's point of view is that the females become progressively less loyal as the size of the harem increases. As a result, males find it proportionately easier to take over larger units because they can count on the females' willingness to desert. As a consequence of these two effects, a male's reproductive lifespan – i.e. that period during which he can breed – is inversely related to the number of females he has in his unit. This, in turn, means that he can breed less often with each of them and, if the rate of decline in his tenure as harem-holder is steep enough, that he will produce fewer offspring over the course of his lifetime than if he had held a smaller unit.

Given that we can determine both how female maturation and mortality rates affect group size and how this in turn affects female reproductive rates and male tenure, we can develop a fairly straight-forward demographic model which allows us to determine how many offspring a harem male can expect to produce over the course of his lifetime if he acquires a harem of a certain size at a given age (for details, see Dunbar, 1984). Two important conclusions emerge from the simulations based on this model. One is that lifetime reproductive output is an inverted U-shaped function of the number of females that the male has at the start of his reproductive career. In other words, the number of offspring he can expect to father over the course of his lifetime increases with the size of his initial harem, but reaches a peak at harem sizes of around six females and then declines again. Males with harems of more than ten females at the start of their breeding careers fare no better than males who have only a single female. This turns out to be a direct consequence of the fact that it is so much easier for males to take over large units. Whereas a male with few females can expect to retain control over them for the whole of his natural lifespan, males with large numbers of females are likely to have very much shorter careers as harem-holders. The other significant finding is that, for any given size of harem, a male's expected lifetime reproductive output declines the older he is when he begins his breeding career. This is obviously a direct consequence of the fact that older males simply have less time remaining to them in which to breed before reaching the natural end of their lives.

These results suggest that a male would father most offspring if he acquired a moderately large unit as early as possible after becoming physiologically capable of breeding. The problem with this ideal solution is that such young males are too small either to wrest control of a unit away from an older incumbent male or to retain control over such a unit once they have done so. The fact that the females live in groups means that there are relatively few harems compared to the number of males in the population that would like to own one. Consequently, the older more powerful males are able to monopolise all the females and younger males are forced to wait on the sidelines until they are old enough to be able to challenge harem-holders with any hope of success. Although males undergo puberty at about four years of age, they do not complete skeletal growth until the age of six years and, even then, they continue to put on weight (and hence increase in fighting ability and power) for at least a further two years. Thus, few males are physically capable of defeating an incumbent harem-holder much before the age of eight.

In most cases where the level, of competition forces large numbers of males to delay the start of their reproductive careers for a significant period of time, Darwinian considerations suggest that these excluded males should search for opportunities that would allow them to circumvent the conventional processes of acquiring mates. Indeed, alternative mating strategies do seem to be a widespread phenomenon throughout the animal kingdom (Dunbar, 1982). Male gelada are no exception and some males are able to acquire access to females in a less fiercely competitive way by joining another male's unit as a 'follower'. Young males can join units in this way by behaving submissively and avoiding all situations that might seem to be challenging the harem-holder's hegemony over the females. An older male, in contrast, would not be tolerated by the harem-holder precisely because the power differential between them is too small and such a male would be in a very good position to spring a successful takeover from within the unit. Once the harem-holder has accepted a young follower into his unit, such a male is able to build up relationships gradually with some of the more peripheral females. In time, these develop into formal breeding relationships, so that the young male and his female(s) begin to behave rather like a unit of their own within the unit. Eventually, they split off and set up as a completely independent unit, but this process takes several years to complete.

Knowing that there are only two key variables that determine a male's lifetime reproductive output (namely, the initial size of his harem and the age at which he begins his breeding career), we can use our demographic model to determine the numbers of offspring that males pursuing each of the two strategies can expect to gain over a lifetime. From empirical data, we know that males pursuing the normal strategy of taking over a complete unit acquire, on average, about five females at about eight years of age, whereas males pursuing the more pacific follower strategy acquire an average of about two females when they are six years old. Interpolating these values into our demographic model reveals that, when all other things are equal, the differences in the parameter values cancel out so that the two strategies yield virtually identical lifetime reproductive outputs.

The problem with these kinds of analyses is that it is not easy to test their validity directly in the field: to do so, we would have to be able to follow an entire cohort of males over the whole of their natural lifespans and, with species as long-lived as primates, this means continuous field studies of 15–20 years. Fortunately, there are ways we can circumvent this problem. One is to use sensitivity analysis to assess the extent to which the model's results depend on the particular constellation of parameter values that we have observed in the field. By systematically varying the slope parameters for each of the nine equations in the model, we find that any change in the values of these parameters drives the ratio of the two strategies' reproductive outputs further away from equality. The only exception occurs if we assume that we underestimated the ages of our animals in the field. Now, this is significant because it has recently been discovered that observers consistently underestimate (and never overestimate) the ages of wild animals whose births they have not witnessed when physical appearance is used as a basis for ageing. In other words, had we aged our animals incorrectly, then our results approach equality of outputs even more closely than they appear to do. Even if this is not the case, however, the fact that the particular constellation of parameter values that we measured in the field yields a close fit to equality suggests that not only is the model quite sensitive to modest errors of measurement but also that the likelihood of producing the results we did by chance alone are quite remote.

An even more powerful way of testing these kinds of models is to use the results of the simulation in conjunction with more general theoretical considerations to make derivative predictions that can be tested with data from the field. If any such predictions can be made at a quantitative rather than a merely qualitative level, then so much the better. I have undertaken two quite separate analyses of this kind to test our models of gelada reproductive strategies, both of which yielded predictions that turned out to be remarkably accurate at a quantitative level. I shall detail only one of these tests here briefly by way of example (for details, see Dunbar, 1984).

Given that the two strategies seem to produce virtually identical lifetime reproductive outputs, evolutionary theory suggests that the strategy set will only be in evolutionarily stable equilibrium if their frequencies are equal. This is exactly what we find. In two separate populations, fifteen out of 27 males and four out of eight males opted for the follower strategy of harem-acquisition. Neither of these two distributions differs significantly from the predicted 50 : 50 ratio.

COGNITIVE ETHOLOGY: A NEW APPROACH?

The emphasis that sociobiology places on the decisions made by animals raises some rather thorny questions about consciousness. Do male gelada, for example, really weigh up the costs and benefits of different strategies of harem acquisition? I am not sure that it matters whether or not we can answer this question, but sociobiology has none the less renewed interest in the question of animal awareness. Cognitive ethology, as it has come to be known, is now a rapidly growing field and I would like to illustrate the kind of work being done in this area with just one example, and then I want to return to Julian Huxley and point out that, more than half a century earlier, he was himself arguing for just such a cognitive approach.

The example that I want to discuss concerns vervet monkeys. Dorothy Cheney and Robert Seyfarth have been studying a population of vervet monkeys in Kenya for some years now and their research has focused increasingly on the question of just what the animals know about their ecological and social environments. Two sets of results are particularly interesting because they tell us something about the knowledge the animals have about their relationships with each other. Note that I stress here not just knowledge about each other but knowledge about their *relationships* with each other.

In an early experiment, Cheney and Seyfarth (1980) recorded the distress calls of particular infants and played these back from a loudspeaker hidden in a bush under very carefully controlled conditions: the infant's mother and at least two other adult females had to be close to the hidden speaker, but the infant itself had to be out of all their fields of view. Now, there are several levels of information that an individual could in principle acquire from hearing the disembodied screams of a particular infant: these are (1) an infant is in distress, (2) the infant in distress is or is not my infant and (3) the infant in distress belongs to that particular female over there. Each of these assumes increasingly greater levels of knowledge about the relationships between the infant and the members of the group. In order to be able to assess this, Cheney and Seyfarth observed the females' responses to playback of the call and compared their directions of gaze during the ten seconds just before and just after the onset of the call. They found that, while the mother invariably spent significantly more time staring at the bush where the loudspeaker was hidden, the other females, if they looked up at all, always looked first at the mother. In other words, they not only recognised the infant's voice as belonging to a specific individual, but they also recognised that that individual bore a particular relationship to a specific mother in the group. Moreover, auditory cues alone were sufficient to trigger recall of that knowledge.

In a later study, Cheney and Seyfarth (1986) attempted to determine the extent to which animals remember past encounters that they have had with other members of their group. Whenever an agonistic encounter occurred, they recorded the behaviour of the protagonists during the subsequent two hours. They found that the loser in the fight was significantly more likely to threaten a relative of its opponent during the two hours after a fight than it did during a control period when no fight had taken place. More significantly, perhaps, unrelated allies of the opponent were also more likely to be threatened following a fight even if the allies had not themselves been involved in that particular encounter.

These findings imply that vervets at least are able to remember the structure of social relationships within a group: not only can they remember whom they have recently been fighting with, but they can also remember whom that individual is related to and which other members of the group have given that individual aid in past fights (and not necessarily fights that involved the current protagonist). Such evidence for quite sophisticated forms of social knowledge has prompted a number of those working with primates to argue that it is not enough simply to describe what animals do. In species as social as the primates (and perhaps many other mammals and birds), we can only begin to understand why animals behave in the way they do if we can work out what it is they are trying to do. For what an animal actually does is a compromise between what it would like to do (its ideal strategy) and what it can do given the constraints imposed on its freedom of choice by the particular biological context within which it happens to live (see, for example, Seyfarth, 1980; Kummer, 1978, 1982; Dunbar, 1984; 1988a). Such a perspective forces us to abandon the behaviourist principles under which we have operated for the past fifty years or so in order to try to enter into the mind of the animal and see the world from the animal's point of view.

Interestingly enough, this was just what Huxley was advocating in the 1910s and 1920s. In his classic paper on grebe courtship, for example, he speaks of trying to understand the bird's psychology in order to gain an insight into the functional aspects of its behaviour (Huxley, 1914). At the heart of this, of course, lay an emphasis on the careful observation of behaviour as our only clue to what is happening inside the animal's mind. Although the ethologists were later to proscribe mentalisms as unscientific under the pervasive influence of behaviourism, Huxley's insistence on the need to see the world from the animal's point of view remained at the very heart of the ethological approach.

CONCLUSIONS

The prominent political role that Julian Huxley played in the establishment and subsequent development of a formal ethological presence in the UK has recently been documented by Durant (1986). In this, Huxley's own enthusiasm for natural history played an important part. I have tried to show that Huxley's influence went far beyond the purely political, for it seems to me that his own field work during the early decades of this century laid the foundations for many of the key features of ethology as we know it today. His studies of the courtship behaviour of water birds were not merely classics of their kind, but they also contained within them the seeds of a surprising number of the concepts that were later developed by the ethologists. Yet, even though the nature of ethology has changed considerably in recent years, I have the impression from reading those early papers that Julian Huxley would very much have approved of the kinds of research now being undertaken by contemporary field ethologists. And I daresay that, were he a young man today, he would be enthusiastically at work at the forefront of those endeavours.

References

Bowman, L. A., Dilley, S. R. and Keverne, E. B. (1978) 'Suppression of Oestrogen-induced LH Surges by Social Subordination in Talapoin Monkeys', *Nature, London*, vol. 275, pp. 56–8.

Cheney, D. L. and Seyfarth, R. M. (1980) 'Vocal Recognition in Freeranging Vervet Monkeys', *Animal Behaviour*, vol. 28, pp. 362–7.

Cheney, D. L. and Seyfarth, R. M. (1986) 'The Recognition of Social Alliances Among Vervet Monkeys', *Animal Behaviour*, vol. 34, pp. 1722–31.

Crook, J. H. (1962) 'The Adaptive Significance of Pair Formation Types in Weaver Birds', *Symposia of the Zoological Society of London*, vol. 8, pp. 57–70.

Crook, J. H. (1965) 'The Adaptive Significance of Avian Social Organisations', *Symposia of the Zoological Society of London*, vol. 14, pp. 181–218.

Crook, J. H. (1966) 'Gelada Baboon Herd Structure and Movement: a Comparative Report', *Symposia of the Zoological Society of London*, vol. 18, pp. 237–58.

Crook, J. H. (1970) 'The Socio-ecology of Primates', in J. H. Crook (ed.), *Social Behaviour in Birds and Mammals* (London: Academic Press) pp. 103–66.

Crook, J. H. and Gartlan, J. S. (1966) 'Evolution of Primate Societies', *Nature, London*, vol. 210, pp. 1200–3.

Dawkins, R. (1976) *The Selfish Gene* (Oxford: Oxford University Press).

Dawkins, R. (1984) 'Universal Darwinism', in D. S. Bender (ed.), *Evolution from Molecules to Men* (Cambridge: Cambridge University Press) pp. 403–25.

Darwin, C. (1872) *The Expression of the Emotions in Man and Animals* (London: Murray).

Dunbar, R. I. M. (1982) 'Intraspecific Variations in Mating Strategy', in P. Klopfer and P. Bateson (eds), *Perspectives in Ethology*, Vol. 5 (New York: Plenum Press) pp. 385–431.

Dunbar, R. I. M. (1984) *Reproductive Decisions: An Economic Analysis of Gelada Baboon Social Strategies* (Princeton, NJ: Princeton University Press).

Dunbar, R. I. M. (1987) 'Habitat Quality, Population Dynamics and Group Composition in Colobus Monkeys (*Colobus guereza*)', *International Journal of Primatology*, vol. 8, pp. 299–330.

Dunbar, R. I. M. (1988a) *Primate Social Systems* (London: Chapman & Hall).

Dunbar, R. I. M. (1988b) 'Social Systems as Optimal Strategy Sets: the Costs and Benefits of Sociality', in V. Standen and R. Foley (eds), *Socioecology of Mammals and Man* (Oxford: Blackwell Scientific Publications) (in press).

Durant, J. R. (1986) 'The Making of Ethology: the Association for the Study of Animal Behaviour, 1936–1986', *Animal Behaviour*, vol. 34, pp. 1601–16.

Emlen, S. T. (1982) 'The Evolution of Helping. I. An Ecological Constraints

Model', *American Naturalist*, vol. 119, pp. 29–39.

Emlen, S. T. and Wrege, P. H. (1986) 'Forced Copulations and Intra-specific Parasitism: Two Costs of Social Living in the White-fronted Bee-eater', *Ethology*, vol. 71, pp. 2–29.

Graffen, A. (1984) 'Natural Selection, Kin Selection and Group Selection', in J. R. Krebs and N. B. Davies (eds) *Behavioural Ecology* (Oxford: Blackwell Scientific Publications) pp. 62–86.

Hamilton, W. J. (1964) 'The Genetical Evolution of Social Behaviour. I, II', *Journal of Theoretical Biology*, vol. 7, pp. 1–52.

Hinde, R. A. (1966) *Animal Behavior* (New York: McGraw-Hill).

Holmes, R. T. (1970) 'Differences in Population Density, Territoriality and Food Supply of Dunlin on Arctic and Subarctic Tundra', in A. Watson (ed.), *Animal Populations in Relation to their Food Resources* (Oxford: Blackwell Scientific Publications) pp. 303–19.

Huxley, J. S. (1914) 'The Courtship Habits of the Great Crested Grebe (*Podiceps cristatus*); with an Addition to the Theory of Sexual Selection', *Proceedings of the Zoological Society of London*, vol. 35, pp. 491–562.

Huxley, J. S. (1923) 'Courtship Activities in the Red-throated Diver (*Colymbus stellatus* Pontopp.); Together with a Discussion of the Evolution of Courtship in Birds', *Journal of the Linnean Society, Zoology*, vol. 35, pp. 253–92.

Huxley, J. S. (1934) 'A Natural Experiment on the Territorial Instinct', *British Birds*, vol. 27, pp. 270–7.

Huxley, J. S. (1942) *Evolution: The Modern Synthesis* (London: Allen & Unwin).

Huxley, J. S. (1959) 'Clades and grades', *Systematics Association Publication*, vol. 3, pp. 21–2.

Huxley, J. S. (1966) 'Introduction [Discussion on Ritualisation of Behaviour in Animals and Man], *Philosophical Transactions of the Royal Society*, Series B, vol. 251, pp. 249–72.

Kummer, H. (1978) 'On the Value of Social Relationships to Non-human Primates: a Heuristic Scheme', *Social Science Information*, vol. 17, pp. 687–705.

Kummer, H. (1982) 'Social Knowledge in Free-ranging Primates', in D. R. Griffen (ed.), *Animal Mind – Human Mind*, (Berlin: Springer) pp. 113–150.

McGrew, W. C. and McLuckie, E. C. (1986) 'Philopatry and Dispersion in the cottontop tamarin, *Saguinus (o.) oedipus*: an Attempted Simulation', *International Journal of Primatology*, vol. 7, pp. 401–22.

Moehlman, P. (1979) 'Jackal Helpers and Pup Survival', *Nature, London*, vol. 277, pp. 382–3.

Myers, J. P., Connors, P. G. and Pitelka, F. A. (1979) 'Territory Size in Wintering Sanderlings: the Effects of Prey Abundance and Intruder Density', *Auk*, vol. 96, pp. 551–61.

Ridpath, M. G. (1972) 'The Tasmanian Native Hen, *Tribonyx mortieri*', *CSIRO Wildlife Research*, vol. 17, pp. 1–118.

Selous, E. (1901) *Bird Watching* (London: Dent).

Seyfarth, R. M. (1980) 'The Distribution of Grooming and Related Behaviours among Adult Female Vervet Monkeys', *Animal Behaviour*, vol. 28, pp. 798–813.

Slater, P. J. B. (1985) *An Introduction to Ethology* (Cambridge: Cambridge University Press).

Stacey, P. B. (1979) 'Habitat Saturation and Communal Breeding in the Acorn Woodpecker', *Animal Behaviour*, vol. 27, pp. 1153–66.

Thayer, G. H. (1909) *Concealing Colouration in the Animal Kingdom* (New York).

Tinbergen, N. (1952) 'Derived Activities: their Causation, Biological Significance, Origin and Emancipation during Evolution', *Quarterly Review of Biology*, vol. 27, pp. 1–32.

Tinbergen, N. (1963) 'On the Aims and Methods of Ethology', *Zeitschrift für Tierpsychologie*, vol. 20, pp. 410–33.

White, G. (1789) *Natural History and Antiquities of Selborne* (London: Benjamin White (reprinted 1974 by Dent: London).

Woolfenden, G. E. (1975) 'Florida Scrub Jay Helpers at the Nest', *Auk*, vol. 92, pp. 1–15.

7 The Problem of the Palaeontological Evidence

T. S. Kemp

INTRODUCTION

The story of how the neodarwinian, or synthetic theory of evolution arose, including the role played by Julian Huxley's book *The Synthetic Theory of Evolution*, has been told many times. Moreover, practically every evolutionary biologist of the past two or three decades has at some stage offered a succinct statement of the synthetic theory reduced to its elemental minimum. Here is yet another variant:

> The basic epistemological structure of the synthetic theory of evolution consists of : (i) the original Darwinian concept of natural selection, whereby the existence of heritable variation and of competition between organisms for resources are given as premises. It follows as a deduction that changes in gene frequencies within a population, and in the phenotypic characters caused by those genes, occur over time, generating increasing fitness. Only in the most trivial, short-term and often artificial of cases are there ever direct empirical observations of natural selection; (ii) Mendelian genetics, and ecological studies which perform the role of testing the two premises of Darwinism to show that they are essentially true of living organisms in general.

Both theoretically, and also from direct studies of living populations, it is clear that significant evolutionary change by neodarwinian means must be extremely slow as measured by the human lifespan, too slow for direct observation. Yet a scientific theory is supposed to be testable, which generally means making observations of nature to see if what the theory predicts should happen, really does so. The fortunate existence of a fossil record, a record however imperfect of former states of organisms, and therefore of real changes over time, offers the only possible means of *directly* testing a theory of evolution.

Given this indisputable potential importance of palaeontology, the question arises of why there is so much debate, controversy and often acrimony associated with the actual interpretation of the fossil record. If the record is simply not good enough to use as a reliable test, then this should be obvious; if it is good enough, then a large measure of agreement would be expected about what it indicates regarding evolutionary processes.

SIMPSON AND SCHINDEWOLF: THE EPISTEMOLOGICAL GAP

The nature of the difficulty facing the interpretation of fossils can be illustrated nicely by a correspondence between two eminent palaeontologists in the 1950s. The philosopher Marjorie Grene wrote an essay in 1958, pointing out that G. G. Simpson and O. W. Schindewolf both claimed that the fossil record supported their respective theories of evolution. Yet their theories could hardly have been more different from one another. Simpson was a neodarwinist, being the person who, in 1944, had incorporated the fossil record into the synthetic theory. He believed that the major cause of evolutionary change was natural selection acting over long periods of time, and producing gradual adaptive change. Schindewolf on the other hand was a typostrophist, believing that major evolutionary changes occurred more or less instantaneously and involved sudden reorganisations of the genome. This produced shifts from one stable morphological state to a new one, dictated by laws of morphological organisation. How could these two authorities look at the same fossil record and yet each claim that it supported his view rather than the other?

What actually happened was that Simpson supported his theory of gradual evolution by supposing that there are large numbers of missing fossils, which would, if found, fill in the gaps between the known forms by imperceptible stages. He 'invented', as it were, missing fossils. On the other hand, Schindewolf supported his typostrophic theory of evolution by reading the fossil record fairly literally, but supposing that there are as yet undiscovered mechanisms for evolutionary change. He 'invented' mechanisms. As Gould (1980) has pointed out with reference to Simpson, so Schindewolf as well demonstrated compatibility between the fossil record on the one hand and a theory of evolutionary mechanisms on the other, by the device of creating additional, uncorroborated assumptions, or *ad hoc* hypotheses.

This story illustrates the general way in which a comprehensive theory of evolution is put together. It is clear that there are two distinct kinds of relevant evidence, with what might be termed an epistemological gap between them (Kemp, 1985). The study of living organisms, such as the nature of their genetic systems, the way that genes affect phenotypic characters, the general phenomena of adaptation, of population structure and so on, lead to the creation of hypotheses about evolutionary mechanisms. This kind of mechanistic or 'process' evidence, has a very high resolution, because successive generations of organisms can be studied. However, it is inevitably very short-term evidence, offering information about evolutionary change over at most a few tens of years and usually a good deal less. The second kind of evidence is taxonomic or 'pattern'. It is concerned with the products of evolution, the different kinds of organisms that have arisen as a result of the process, and it includes palaeontological studies. In contrast to mechanistic evidence, palaeontological evidence is of low resolution because the nature of stratigraphy is such that a given fossil specimen cannot be dated relative to another closer than, at best some thousands of years and usually very much more. On the other hand, such evidence has a very long time span, covering the whole of the fossil record.

A comprehensive theory of evolution must involve both kinds of evidence. Its purpose is to propose a mechanism of evolutionary change which generated the actual taxonomic diversity of organisms that exists and has existed in nature. As with scientific theories in general, the evolutionary theory is about proposed causes (processes) that produced perceived effects (patterns) in nature. As such, there is one property of the theory which is the *sine qua non* for its possible acceptance. This is that there exists compatibility between the proposed cause and the perceived effect: the one must be capable in principle of generating the other. When such a state of compatibility does not immediately exist, then it must be created by the use of *ad hoc* hypotheses. These additional assumptions have been neither corroborated nor refuted by empirical observations, and may therefore be true. Equally, they may be false.

In this way, Simpson found that his proposed neodarwinian, gradualistic evolutionary process was incompatible with the observation that the fossil record did not contain long continuous sequences of gradually changing fossils. He defended the neodarwinism by the *ad hoc* assumption that the fossil record is highly incomplete. Obviously there is no direct empirical test possible that shows just what the supposed missing fossils are like, even if it is demonstrated that the stratigraphy is incomplete. In like fashion, Schindewolf also in effect noted incompatibility between the highly punctuated, saltatory fossil record and the neodarwinian mechanism. He accordingly defended his rather literal interpretation of the fossil record by the *ad hoc* assumption that a typostrophic mechanism for saltatory evolutionary change exists. As with Simpson's fossils, so a failure to discover a mechanism for typostrophic change does not refute it, at least until knowledge of the genetic basis of evolutionary change is complete.

In principle, the way to decide objectively which of two competing theories is the preferable is to decide which of the two incorporates the least amount of *ad hoc* assumption. In simple terms this is because an *ad hoc* assumption is part of a theory that may be untrue. Therefore of two theories which both offer an explanation of the same phenomena, the one with the greater *ad hoc* content is the one with the greater statistical probability of being overall false; it has a larger capacity for being wrong, so to speak. Of course there are profound difficulties in the way of getting realistic measures of the *ad hoc* components of theories, which is precisely why there remain arguments about which theory is preferable, particularly in a complex area like evolutionary theory. Are Simpson's missing fossils more, or less *ad hoc* than Schindewolf's undiscovered mechanism? Nevertheless, the principle is clear.

What has frequently happened in the history of evolutionary studies, and indeed remains the *modus operandi* of many evolutionary biologists, is that an entirely subjective decision is made about whether the pattern evidence or the process evidence is to be taken as paramount. With reason if not excuse, students of population genetics have tended to regard their processes as sufficient to refute an apparently inconsistent fossil record. Contrariwise the palaeontologists have all too often accepted that the fossil record is sufficiently complete to show that the mechanisms proposed by population geneticists must be inadequate explanations for evolutionary change. Gould (1983), in his perceptive essay on the history of palaeontology, has described an early phase of independence, when palaeontology created its own theories with little reference to neontological studies.

Then with the coming of the neodarwinian synthesis, palaeontology was very much subordinated to the neontological view, and indeed to the mathematical modelling of population genetics. As exemplified by Simpson, it was the fossil evidence that was normally the subject of *ad hoc* assumptions, not the ideas of mechanism. Gould then speaks of the current situation as a partnership between palaeontology and neontology, which ideally is the situation just outlined here. Whether this is indeed the case may now be illustrated with a contemporary problem, the process by which new species arise.

THE PROBLEM OF SPECIATION

By definition, once a species has split into two, an irreversible phylogenetic step has occurred. Speciation therefore plays a fundamentally important role in evolution, and an account of the mechanism involved is critical for evolutionary theory. Yet, despite the enormous effort made, there is still no agreement about the details of exactly how speciation occurs. Basically there are two conflicting general kinds of theory, with several variations of each, which have been reviewed recently by Carson and Templeton (1984), and Barton and Charlesworth (1984). Briefly they may be categorised as follows:

(i) *The strictly neodarwinian theory.* Speciation occurs as an outcome solely of the neodarwinian processes, particularly natural selection but also possibly a degree of genetic drift, acting differently upon separate populations of a species. Simply after enough time has passed, enough genetic divergence will have occurred for the respective populations to have ceased to be genetically compatible. They will have become unable to produce viable hybrids, and are therefore separate species. The basis for this theory is partly an extrapolation from empirical observations on selection in living populations, both natural (e.g. Endler, 1986) and artificial. It can also be deduced from the genetic structure of populations, particularly with the help of mathematical modelling on the basis of what are taken to be realistic assumptions.

(ii) *Genetic revolution theory* Speciation is a process that has to be triggered by an accidental, and therefore a non-adaptive event. A well established species is seen as highly resistant to evolutionary change because of what is often termed, rather loosely, genetic homeostasis. The integration of the genes forming the overall gene pool, and their complex epistatic interactions see to it that new mutations are generally not capable of being incorporated into the gene pool, except for those with trivial effects. Formation of a new species can only occur as a result of what is termed a genetic revolution, whereby the homeostatic system is broken down, to be followed by the rapid evolution of a new homeostatic system, under the influence of selection and genetic drift. The initial cause of such a revolution is usually postulated to be the founder effect, in which a very small part of the population, carrying a very biased sample of the gene pool, becomes isolated. The evidence adduced to support a genetic revolution theory includes the claim that in the case of those modern species which range over a wide variety of environmental conditions, there is much less geographic variation than would be expected given the presumed different selection forces acting in different parts of the species' range. This is taken to suggest a lack of efficacy of natural selection in a large population. There are also observations about certain taxonomic groups of organisms where many more species are found to be living in archipelagos than are found living on continental land masses. The most intensively studied case of this concerns the genus *Drosophila*, where the Hawaiian Islands support some hundreds of different species. This is taken to be evidence that under circumstances where founder events are inherently likely to be frequent, speciation events are also commoner.

There are also attempts to model speciation along the lines of this kind of theory, again making what are assumed to be realistic assumptions about population genetic structure and the likely behaviour of genes under founder conditions.

Obviously these two respective kinds of theory of speciation make different predictions about the course of history of species. They are therefore at least potentially testable by the fossil record, which introduces the familiar, continuing debate between the gradualistic and the punctuated equilibria views of evolution. In their 1972 paper, Eldredge and Gould pointed out that most fossil species appeared abruptly in the fossil record, rather than being connected to ancestral species by a series of preserved intermediate forms. Species also typically remain in the record for a long period, some millions of years, without major change, and eventually go extinct as abruptly as they arose. This is the familiar punctuated equilibrium pattern of fossils, and is interpreted by many people as evidence for a genetic revolution kind of mechanism of speciation. The long periods of stasis are taken to indicate the existence of genetic homeostasis causing the population to resist any selective forces present that tend to promote changes in the species over time. The punctuation events are interpreted as rapid morphological change, consequent upon the breaking up of the genetic homeostasis, at least possibly by a founder event that is invisible in the record. It should be stressed that this general interpretation of an apparently punctuated equilibrium fossil record is not necessarily a denial of any role for natural selection in morphological evolution. What it does imply is that natural selection requires particular circumstances in order to be effective in promoting change. A population reduction, for example, after a founder event or perhaps an extreme environmental perturbation, promoting a genetic revolution, would be followed by rapid adaptive adjustments.

There has continued to be a lively, unresolved debate on two interrelated fronts. First, do most fossil species actually show this pattern, and second, if so, is the punctuated equilibrium pattern inconsistent with the strictly neodarwinian theory of evolution in general and of speciation in particular?

THE CASE OF THE LAKE TURKANA MOLLUSCS

Still the most celebrated fossil record which is claimed to show the punctuated equilibrium pattern is the one described by Williamson (1981a). It consists of several species of freshwater molluscs, about thirteen of which are preserved in sufficient abundance for statistical analysis, occurring over some 1000 sq km to the east of Lake Turkana in East Africa. They occupy about 400 metres of plio-pleistocene deposits, which represents approximately 4 million years of real time. Since the publication of Williamson's initial paper, several authors have commented upon the meaning to be drawn from these fossils, and a number of mutually incompatible hypotheses have been proposed. This case makes a very good example for illustrating the nature of contemporary palaeontological disputes, because the different hypotheses can be shown to differ most fundamentally in the *ad hoc* assumptions that their respective proponents have chosen to make.

The Williamson theory

Williamson (1981a) observed three significant aspects of the overall pattern of the fossils (Figure 6.1). First, for most of the period covered all the species showed stasis, with no evolutionary change detectable. Indeed, stasis occupied 99 per cent or more of the total time. Second, on a small number of occasions a series of new species appeared in place of the existing, apparently ancestral species. At the temporal resolution available, several speciation events were simultaneous with one another, and allopatric in that the ancestral species did not coexist with the respective descendant species. The new species were also temporary in that they shortly disappeared, being replaced by reappearance of the original species, which must have been surviving elsewhere. The speciation events also appear to be correlated with regressions of the lake system of the time, that is, temporary falls in lake level. The third observation by Williamson was that in a period estimated at between 5000 and 50 000 years immediately prior to the appearance of a new species, there was a marked increase in the phenetic variability of the ancestral species population. This he regarded as showing that some kind of genetic revolution had been in progress, manifesting itself as a breakdown in the developmental mechanism of the organisms.

Thus Williamson concluded that the Lake Turkana fossil mollusc record is sufficiently complete, the samples sufficiently close to real life samples, and the stratigraphic analysis sufficiently accurate, for it to be taken as an adequate direct record of what really happened. As such, it corroborates a genetic revolution kind of theory of speciation, with stasis the norm, and occasional very rapid speciation initiated by something that caused a genetic revolution. That something was correlated with a major environmental change, and was manifested by a temporary breakdown of the mechanisms maintaining a constant developmental pattern within the members of the species. The only feature of this account strikingly at odds with existing ideas of genetic revolution speciation is that there is no direct evidence for a founder effect, or a bottleneck in the population. The whole population undergoes the increased variability. It would, however, be a simple matter to create an *ad hoc* assumption that a drastic population reduction did occur, but that it lasted for such a brief time that it is below the resolution of the fossil record, and therefore undetectable.

No sooner had Williamson's (1981a) account appeared than it

88

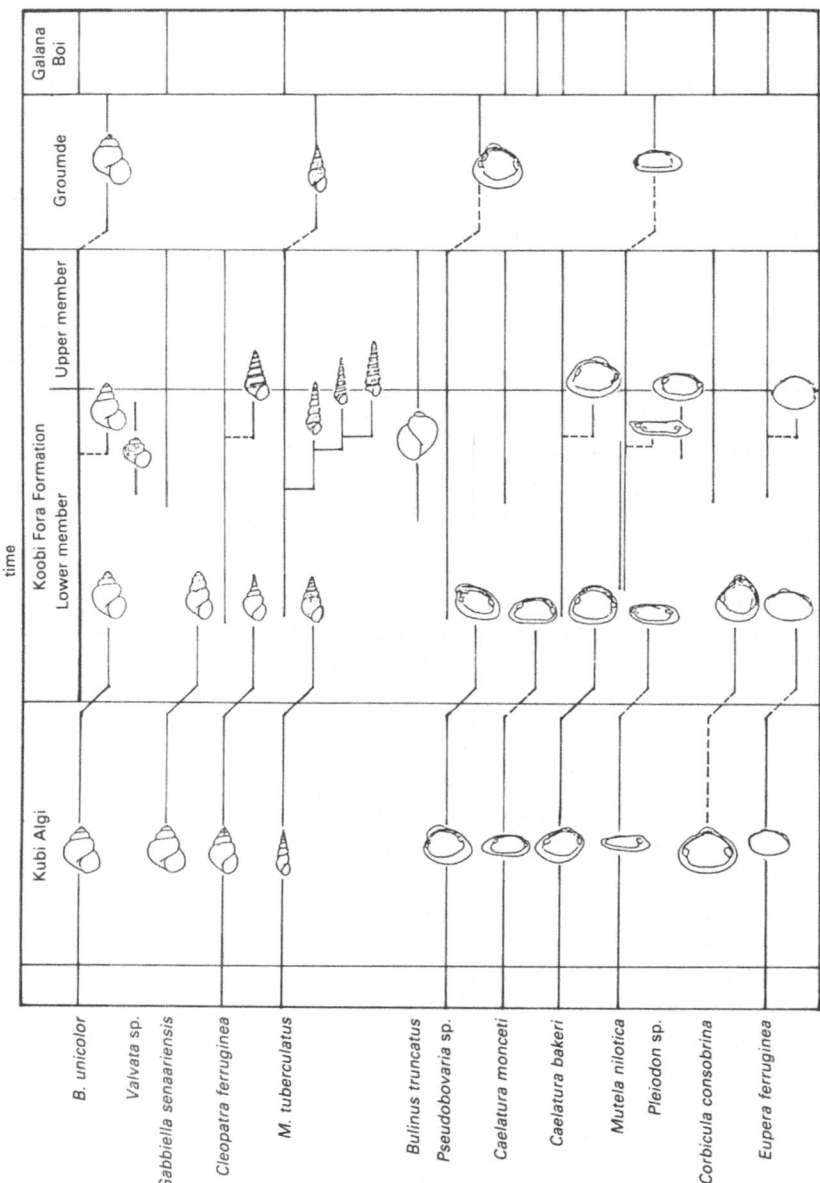

Figure 7.1   The fossil record of the main species of mollusc in the Turkana Basin. Somewhat simplified from Williamson (1981a).

faced several challenges, and indeed continues to do so. The two major alternatives on offer are either that the pattern seen is perfectly compatible with a strictly neodarwinistic, totally selective mechanism, or that no evolution had actually taken place, all the apparent change being due to ecophenotypic (that is, non-genetic, environmentally induced) change, or alternatively due to the immigration of pre-existing species from elsewhere.

The strict neodarwinist explanation

Several authors have proposed that the Lake Turkana mollusc record can be adequately explained by the conventional mechanism of neodarwinian selection (Jones, 1981; Maynard Smith, 1982; Charlesworth and Lande, 1982). The periods of stasis were due to stabilising selection, whereby the existing morphology had the highest fitness and was consequently favoured by natural selection for as long as the environment remained unchanged. The rapid shifts in morphology resulting in the new species were caused by periods of intense selection, occurring when the environment did change.

The rates of evolutionary change implied by this explanation are perfectly consistent with rates of change known for modern populations. Charlesworth and Lande (1982) argued explicitly that modern mollusc species show heritable variation, and if a directional selection force is present, they must respond by evolutionary change. Assuming the same to be true of the fossil populations, absence of change therefore implies absence of directional selection forces. These same authors also comment on the increased variability noted prior to the establishment of the new species. Rather than allowing Williamson's (1981a; 1981b) rather vague explanation that it results from the breaking down of genetic homeostasis, they suggest that it might be due to a temporary heterogeneity of the environment, or perhaps an artefact arising from the lumping together as a single sample the members of a rapidly evolving lineage.

Jones (1981) quotes a number of examples where artificially applied selection has produced morphological change in a population far exceeding the rates implied by the Lake Turkana molluscs, and therefore he sees no reason to abandon rapid directional selection as the sole explanation for the origin of new species in this particular case.

A strictly selectionist, neodarwinian explanation of this kind entails some of the same *ad hoc* assumptions as Williamson's theory, namely that the morphological changes observed represent genetic shifts in the populations, and that the putative new species were indeed true species, reproductively isolated from their parent species. However, it differs in making certain other equally untestable assumptions. The explanation offered for stasis implies that for very long periods, of at least four million years for most species, the environment remained sufficiently constant that there were no directional selection pressures promoting evolutionary change. Obviously this statement applies to shell form directly, but also, given the assumption that the distinct morphological forms are true species, then it follows that most of the unpreserved characters must also have been maintained static by stabilising selection in a constant environment. As far as the speciation events are concerned, a rather general *ad hoc* assumption must also be made. This is that the mechanism underlying artificial selection can be extrapolated to apply to natural selection both over a far longer period, and also within the vastly more complex circumstances of a natural environment as opposed to the much more controlled nature of laboratory and domestic environments.

Charlesworth and Lande's (1982) explanations for the increased variability require further *ad hoc* assumptions, to the effect that spatial heterogeneity of an environment can result in the degree of variation seen, or that the stratigraphic resolution is less than claimed by Williamson. Williamson (1982) has indeed challenged both assumptions explicitly, the first by claiming that modern populations of molluscs do not show such a degree of spatial variation, and the second by pointing out that amongst the variants are forms which are not actually intermediate in structure between the parent and the descendant species in question.

The ecophenotypic explanation

Mayr (1982) and Boucot (1982) both briefly suggested that the morphological changes observed might represent non-genetic change in the population caused by the direct effect of environmental change on the development of individual organisms, an ecophenotypic effect. Kat and Davis (1983) discussed this possibility in greater detail, pointing to examples amongst modern molluscs comparable to the Lake Turkana species, where they claimed that ecophenotypic variation matches the extent of the differences attributed by Williamson (1981a) to different species. They also suggested that the appearance of phenotypic change in many lineages simultaneously, including some supposedly asexual forms, supports an ecophenotypic rather than an evolutionary explanation.

Williamson (1982; 1983) countered their view by the comments that the variant forms co-exist for a time with the ancestral forms; that adequate morphometric analysis shows that much less pronounced ecophenotypic effects occur in the widespread modern populations of these molluscs; and that achievement of the new forms took some 1000 to 10 000 years, which is hardly the time span expected of a phenotypic effect.

Nevertheless, Fryer *et al.* (1983; 1985) developed the idea of an ecophenotypic explanation in detail, with consequent rebuttals by Williamson (1985a; 1985b). The gist of the argument is the claim by Fryer *et al.* that variation due to such environmental factors as differences in salinity, temperature, pH etc. in modern freshwater mollusc species is adequate to explain the novel forms found in the Lake Turkana deposits. Williamson denies that this is the case. Both quote literature and observations supporting their respective standpoints, and the matter is not yet resolved.

At any event, the *ad hoc* assumptions underlying an ecophenotypic explanation are fairly extensive. It has to be assumed that the morphological changes occurring within the lineages had no genetic basis, an assumption that can be corroborated to the extent that the degree of ecophenotypic variation in similar species of modern molluscs matches that of the fossil forms. Like in the neodarwinian explanation, it must also be assumed that the environment remained stable for very long periods, with just the right kind of perturbations occurring to cause the brief periods of altered morphology, followed by reversions to the norm. It would also seem necessary to assume that the increased variation occurring in the populations of molluscs prior to the establishment of the new morphology is an indication that this is, to say the least, a very complex case of ecophenotypy.

The ecological displacement theory

Cohen and Schwartz (1983) suggested that the appearance in the record of the new species does not indicate speciation events, but merely the invasion into the area by deeper water species, during brief periods of increase in the level of the lake, a theory given rather short shrift by Williamson (1983). For example, it requires that *all* the species of deep water molluscs were completely excluded from shallower water, otherwise they should appear at least sporadically throughout the record. It also fails to account for the existence of the variant morphologies prior to the supposed invasion. Thus acceptance of such an theory requires the creation of rather complex *ad hoc* assumptions in order to explain away these difficulties.

Comments

It is not the purpose of this short essay to attempt to decide which of these explanatory theories of the cause of the Lake Turkana fossil mollusc pattern is the best, for that would require a great deal more detailed study of the actual material, and of the general properties of freshwater mollusc species. Rather, it is an attempt to illustrate the way in which palaeontological evidence should be used in the context of evolutionary problems, illustrated here by the question of speciation.

One possibility, encompassed by both the ecophenotypic theory and the ecological displacement theory, is that this particular fossil record is irrelevant to the question of speciation, because no speciation events are actually recorded. However the record can also be assumed to illustrate speciation, and can be interpreted in a way that is compatible with either a genetic revolution kind of explanation, or with a strictly neodarwinian, selectionist theory. These various theories do not differ in whether they have been corroborated or refuted by the fossil evidence, but only in the extent and the nature of the necessary *ad hoc* assumptions that they respectively entail. Any one of them (or combination of more than one, for that matter) may be true. The scientific judgement of which theory is the preferable should lie with an assessment of their relative *ad hoc* information contents. That with the least is, by standard scientific reasoning, the best.

The various authors of these various interpretations have tended to approach the problem in a spirit of advocacy based on prejudice about which empirical observations are to be taken as paramount. Williamson (1981a; 1985a) as a palaeontologist primarily defends his fossils; Charlesworth and Lande (1982) as theoretical evolutionary geneticists defend their models of neodarwinism; Fryer *et al.* (1983; 1985) remain steadfastly committed to their knowledge and study of ecophenotypy in living populations. What is needed in the first instance are explicit statements of, and attempts to assess the relative amounts of *ad hoc* information incorporated in the explanations. This would allow a rational rather than an emotive basis for preference; it would also clarify those areas in which further empirical knowledge about the fossils, about the lithology of the sediments, and about the biological nature of living organisms might lead to further refinement of the choice of theory to be accepted.

CONCLUSION

As a contribution to the problem of species and speciation, palaeontology has seemed depressingly agnostic; even the very best example of a punctuated equilibrium pattern of fossils yet discovered seems to generate more controversy than enlightenment, and fails to convince many evolutionary biologists of its value for testing theories. But the fossils do indisputably exist, and cannot therefore legitimately be ignored by any theory. The great problem is still the epistemological gap between the short-term, high resolution knowledge about the nature of living organisms on the one hand, and the long-term, low resolution knowledge about fossil patterns on the other. Not all evolutionary biologists can yet be said to have really appreciated the enormity of this gap. To extrapolate from ecological observations to events seen in the fossil record that took tens and hundreds of thousands of years to come to fruition requires a huge act of faith on the part of neontologists. On the other hand, that everything a fossil was and did simply must conform with general properties of living organisms should remain in the mind of every palaeontologist.

This essay will have served its purpose if it has helped to clarify the problem of the relationship between fossil and neontological evidence, and has shown that there is a proper scientific procedure for relating the two. The best theory will always be the one which combines all the empirical knowledge with the minimum possible amount of *ad hoc* assumptions, irrespective of exactly what the *ad hoc* assumptions actually are. A priori, no particular set of observations or beliefs has paramountcy over the rest.

References

Barton, N.H. and Charlesworth, B. (1984) 'Genetic Revolutions, Founder Effects, and Speciation', *Annual Review of Ecology and Systematics*, vol. 15, pp. 133–64.
Boucot, A.J. (1982) 'Punctuationism and Darwinism Reconciled? The Lake

Turkana Mollusc Sequence. Ecophenotypic or Genotypic?', *Nature, London*, vol. 296, pp. 609–610.

Carson, H.L. and Templeton, A.R. (1984) 'Genetic Revolutions in Relation to Speciation Phenomena: the Founding of New Populations', *Annual Review of Ecology and Systematics*, vol. 15, pp. 97–131.

Charlesworth, B. and Lande, R. (1982) 'Punctuationism and Darwinism Reconciled? The Lake Turkana Mollusc Sequence. Morphological Stasis and Developmental Constraint: No Problem for neo- Darwinism', *Nature, London*, vol. 296, p. 610.

Cohen, A.S. and Schwartz, H.L. (1983) 'Speciation in Molluscs from Turkana Basin', *Nature, London*, vol. 304, pp. 659–60.

Eldredge, N. and Gould, S.J. (1972) ' Punctuated Equilibria: an Alternative to Phyletic Gradualism', in T.J.M. Schopf (ed.), *Models in Paleobiology*, (San Francisco: Freeman, Cooper and Co.) pp. 82–115.

Endler, J.A. (1986) *Natural Selection in the Wild* (Princeton: Princeton University Press).

Fryer, G., Greenwood, P.H. and Peake, J.F. (1983) 'Punctuated Equilibria, Morphological Stasis and the Palaeontological Documentation of Speciation: a Biological Appraisal of a Case History in an African Lake', *Biological Journal of the Linnean Society*, vol. 20, pp. 195–205.

Fryer, G., Greenwood, P.H. and Peake, J.F. (1985) 'The Demonstration of Speciation in Fossil Molluscs and Living Fishes', *Biological Journal of the Linnean Society*, vol. 26, pp. 325–36.

Gould, S.J. (1980) 'G.G. Simpson, Paleontology and the Modern Synthesis', in E. Mayr and W.B. Provine (eds), *The Evolutionary Synthesis* (Harvard: Harvard University Press) pp. 153–72.

Gould, S.J. (1983) 'Irrelevance, Submission and Partnership: the Changing Role of Palaeontology in Darwin's Three Centennials and a Modest Proposal for Macroevolution', in D.S. Bendall (ed.) *Evolution from Molecules to Man* (Cambridge: Cambridge University Press) pp. 347–66.

Grene, M. (1958) 'Two Evolutionary Theories', *British Journal of the Philosophy of Science*, vol. 9, pp. 110–27 and 185–93.

Jones, J.S. (1981) 'An Uncensored Page of Fossil History', *Nature, London*, vol. 293, pp. 427–8.

Kat, P.W. and Davis, G.M. (1983) 'Speciation in Molluscs from Turkana Basin', *Nature, London*, vol. 304, pp. 360–1.

Kemp, T.S. (1985) 'Synapsid Reptiles and the Origin of Higher Taxa', *Special Papers in Palaeontology*, No. 33, pp. 175–84.

Maynard Smith, J. (1982) 'Evolution – Sudden or Gradual', in J. Maynard Smith (ed.), *Evolution Now* (London: Macmillan) pp. 125–8.

Mayr, E. (1982) 'Punctuationism and Darwinism Reconciled? The Lake Turkana Mollusc Sequence. Questions Concerning Speciation', *Nature, London*, vol. 296, p. 609.

Simpson, G.G. (1944) *Tempo and Mode in Evolution* (New York: Columbia University Press).

Williamson, P.G. (1981a) 'Palaeontological Documentation of Speciation in Cenozoic Molluscs from Turkana Basin', *Nature, London*, vol. 293, pp. 437–43.

Williamson, P.G. (1981b) 'Morphological Stasis and Developmental Constraint: Real Problems for Neo- Darwinism', *Nature, London*, vol. 294, pp. 214–15.

Williamson, P.G. (1982) 'Williamson Replies', *Nature, London*, vol. 296, pp. 611–12.

Williamson, P.G. (1983) 'Speciation in Molluscs from Turkana Basin', *Nature, London*, vol. 304, pp. 661–3.

Williamson, P.G. (1985a) 'Punctuated Equilibrium, Morphological Stasis and the Palaeontological Documentation of Speciation: a Reply to Fryer, Greenwood and Peake's critique of the Turkana Basin Mollusc Sequence', *Biological Journal of the Linnean Society*, vol. 26, pp. 307–24.

Williamson, P.G. (1985b) 'In reply to Fryer, Greenwood and Peake', *Biological Journal of the Linnean Society*, vol. 26, pp. 337–40.

8 Size, Shape and Evolution

Robert D. Martin

INTRODUCTION

Given the enormous scope of Julian Huxley's research interests, most modern biologists are bound to acknowledge some kind of intellectual debt to his contributions. When it comes to the question of *size relationships* in organisms, however, a double debt to Huxley must be recognised. In a very general context, there is of course Huxley's pervasive influence in the sphere of evolutionary biology, which constitutes the overall framework within which biological size relationships can be interpreted. To take just one example, he provided a timely and valuable discussion of the concept of 'grade' with respect to evolutionary processes and taxonomy (Huxley, 1958). As it happens, the grade concept has also proved to be of particular relevance with respect to adjustments to body size in the evolution of organisms (see below). More specifically, studies of the relationships between the size of individual bodily components and overall body size (scaling biology) can be said to owe much of their origin to Huxley's seminal treatise *Problems of Relative Growth* (1932). When it first appeared, this book provoked a flurry of discussion and active practical application that lasted for about two decades. Thereafter, interest subsided quite markedly as if a once promising mine had eventually been exhausted; but – following a steady resurgence – interest in scaling biology has peaked again in recent years and studies in this area have now become something of a 'growth industry'.

Scaling studies have been particularly emphasised in comparative vertebrate physiology for some time and Schmidt-Nielsen (1984) has recently provided a very effective review of some of the most striking and successful applications in this field. Following a very early start with respect to the study of relative brain size (Snell, 1891; Dubois, 1897; for a review see Jerison, 1973) the central importance of scaling effects has been slowly but progressively acknowledged in the field of comparative morphology. To take one example, a recent edited volume has brought together a variety of specific applications in morphological aspects of primate evolutionary biology (Jungers, 1985). In recent years, scaling studies have also been extended into essentially new areas, such as comparative ecology (Peters, 1983) and the analysis of life history parameters (Calder, 1984). Huxley would undoubtedly have been pleased to see the extension of scaling studies from his original context of developmental biology into the realms of behaviour and ecology, which later figured so prominently in his research. Special mention should also be made of the entertaining and instructive book by McMahon and Bonner (1983), which neatly demonstrates the generality of scaling effects in biology and graphically shows that the topic is both thriving and productive.

One important point that must be made at once is that it is necessary to distinguish clearly between *interspecific* and *intraspecific* scaling. Huxley's concern with scaling biology was essentially confined to the developmental processes involved in the growth of individuals and in his influential book (1932) he was hence primarily interested in intraspecific scaling (hence the title 'relative growth'). Only one chapter of the book was devoted to the implications of scaling in phylogenetic, rather than ontogenetic, terms. By contrast, in the modern resurgence of scaling biology, interspecific aspects have been dominant and this may explain the curious hiatus in the history of this subject. The temporary decline in interest in scaling as a topic for research may be attributed to the fact that its immediate contribution to the understanding of individual development within species was somewhat limited, while it was only gradually realised that it had an absolutely fundamental part to play in the understanding of phylogenetic relationships between species.

It should also be noted that nonlinear scaling of individual bodily components to body size is now generally referred to as *allometry* (Huxley and Teissier, 1936; Reeve and Huxley, 1945). Initially, Huxley (1932) did not use the word allometry at all and employed instead (following Pézard, 1918) the term *heterogony*. This term was subsequently rejected by Huxley and Teissier (1936), in a brief paper devoted to the terminology of relative growth, because 'heterogony' had also been employed to refer to a particular kind of life cycle. This joint publication of Huxley and Teissier is of special interest for two reasons. In the first place, it provided a valuable early definition of many of the basic terms and concepts of allometric analysis. In addition, however, it rightly associated Teissier's name with a crucial stage in the development of allometric analysis. Although it is now common practice to acknowledge Huxley's seminal contributions to allometry, Teissier's equally significant contributions (e.g. Teissier, 1931; 1948; 1955; 1960) are commonly overlooked.

It must also be emphasised that there are different kinds of intraspecific and interspecific allometry (Gould, 1966). Intraspecifically, one might follow Huxley's example in examining scaling relationships in the course of individual development (*ontogenetic allometry*); but it is also possible to study the way in which individual parameters scale to body weight among adults of different body sizes (*static adult allometry*). With interspecific allometry, on the other hand, most studies are concerned with the scaling of individual parameters to body size across a range of contemporaneous species (*static interspecific allometry*), although it is in principle also possible to examine scaling through a series of fossil forms (*evolutionary allometry* or *phylogenetic allometry*). (In fact, the latter concept is suspect, as it confounds evolutionary changes in body size with evolutionary changes in patterns of bodily organisation (Martin, 1980). As these are the very two components that scaling analysis is designed to separate (see below), it might be preferable to discreetly forget the concept of phylogenetic allometry). Although it is possible to apply the standard scaling formula (see below) to all of the different kinds of allometry listed above, it is important to recognise that the underlying assumptions and problems are to a large extent distinct for any particular context.

In his studies of intraspecific scaling, Huxley examined a considerable variety of organisms and most of the examples in *Problems of Relative Growth* were drawn from invertebrates. It should, however, be noted that he used the example of development of the skull in the chacma baboon (*Papio ursinus*) to illustrate conspicuous changes in shape with changes in size. This example, which involves extremely rapid growth of the muzzle in contrast to the braincase, was subsequently studied in greater detail by Freedman (1962). In this connection, it should also be mentioned that Huxley's ideas had been clearly influenced by D'Arcy Thompson's ideas on changes in shape in the course of growth (Thompson, 1917). He provided a number of illustrations exemplifying D'Arcy Thompson's use of the graphic method to reveal the effects of differential growth gradients (Cartesian transformation). One particularly striking illustration concerns changes in shape of the skull during the evolution of the horse. These examples illustrate the fact that, with respect to morphology, scaling biology can often be seen as the study of change in *shape* with changing size both in the course of individual development and in the evolution of species. Indeed, it is commonly stated that the central problem of scaling analysis is the separation of 'shape' and 'size'. This view is, however, unjustifiable for at least two reasons. In the first place, there are many examples in which scaling analysis is not concerned directly with physical form, but with other biological properties such as physiological parameters (e.g. basal metabolic rate) or temporal components of reproduction (e.g. gestation period). Secondly, in many cases the nature of the scaling relationship is such that shape must in fact change in a progressive and predictable fashion with body size. In the scaling of the mammalian skull, for instance, there are general shape changes that occur with increasing body size because individual components of the skull become relatively larger or smaller. It is therefore misleading to think in terms of 'shape' as some independent component that must be separated from size. The key point with interspecific scaling analysis is to recognise regular trends in biological properties with body size and to identify those species that depart from the general tendency. Given these considerations, it is hence advisable to adopt a more general definition of scaling, such as that of Gould (1966), who defined allometry simply as 'the study of size and its consequences'.

In the context of evolutionary biology, scaling analysis (*allometric analysis*) can now be seen as a very valuable tool that greatly facilitates the task of unravelling phylogenetic relationships. Indeed, it can be said that such analysis is often *essential* for successful interpretation of evolutionary relationships, as body size has far-reaching effects throughout the organisational biology of organisms. The aim of this present contribution is to set out some currently recognised basic principles of allometric analysis, with due reference to the influence exerted by Huxley, and to illustrate the range of practical applications. It should be emphasised that the concern will be very much with *interspecific allometry*, as it is in this area that scaling analysis has recently made the greatest contribution to our understanding of phylogenetic relationships. In addition, the examples will be taken largely from primates. In this respect the author's interests happen to coincide more closely with those of Huxley's equally illustrious forebear Thomas (see Huxley, 1864).

INTERSPECIFIC SCALING

The basic problem to be tackled with scaling analysis can be aptly illustrated with respect to the great range of body sizes found among the living primates. There is an approximately 2000-fold range of body weights extending from the smallest primates such as the lesser mouse lemur (*Microcebus murinus*) and the dwarf bushbaby (*Galago demidovii*), with body weights of approximately 60 g, up to the male gorilla (*Gorilla gorilla*), with a body weight of approximately 120 kg. When any attempt is made to reconstruct phylogenetic relationships among the various primate species, the differences in body sizes immediately present a problem. Given a comparison between any two species of different body sizes, it is necessary to ask the question: 'To what extent do these two species differ merely because they differ in body size and to what extent do they differ because of some fundamental difference in biological organisation.' This is to some extent a rhetorical question, as change in body size during evolution is subject to the same process of natural selection as are other aspects of an organism's biology. Nevertheless, it is reasonable to propose that, under selection, a change in body size may be brought about far more easily than any kind of fundamental reorganisation, as it requires no more than arrest or extension of existing growth programmes. This is, for instance illustrated by the 35-fold range in body sizes brought about by artificial selection applied to the domestic dog (Kirkwood, 1985). It has been relatively easy to produce dogs of different body sizes through artificial selection, but it would seem that no fundamental biological reorganisation has been achieved. Although shape changes do occur with increasing body size in domestic dogs, these are essentially restricted to predictable scaling effects. The key point is that all dogs may be said to share a common growth programme and that differences between breeds arise essentially because of differences in adult body size.

Returning now to interspecific comparisons, it may be proposed that closely related species of different body sizes will also share a common growth programme and that differences in shape between them are similarly attributable entirely to scaling effects. This has been demonstrated, for example, for the African apes (chimpanzee and gorilla) by Shea (1981, 1985). In many respects, it can be stated that gorillas are simply chimpanzees that have grown to attain a larger adult body size. By contrast, it is likely that distantly related species will differ not only in body size but also in their growth programmes. To put it another way, distantly related species will differ significantly even when there is no difference in body size. The aim of allometric analysis in the context of phylogenetic reconstruction, therefore, is to separate differences that are due merely to body size from differences that reflect some underlying reorganisation of growth programmes and their consequences.

The allometric formula

At present, allometric analysis is predominantly an empirical tool. A given bivariate data set is analysed in the attempt to eliminate the general influence of size from the pattern of differences between species. The data set consists of paired values for a given biological parameter (Y) and for a measure of body size (X), taking average values for all species in the sample. Following the early work of Huxley (1932) and using the notation proposed by Huxley and Teissier (1936), the approach is founded upon the standard allometric formula:

$$Y = b.X^{\alpha}$$

The two constants in the equation, the allometric exponent (α) and the allometric coefficient (b), reflect the specific nature of the scaling relationship in any given case. Huxley and Teissier (1936) referred to the exponent α as the *equilibrium constant* and to the constant b as the *initial growth-index*. These terms were, however, specifically tailored for the context of individual growth and are not suitable for general use.

The allometric formula is usually converted into logarithmic form, as it then becomes linear and amenable to simple statistical treatment:

$$\log Y = \alpha. \log X + \log b$$

By determining an appropriate best-fit line for a logarithmically-converted set of data, the values of α and b can be established empirically for a given scaling relationship.

It is possible to define a general biological model for scaling relationships in organisms (Figure 8.1), taking the relationship between some appropriate dimension of a selected character (e.g. the weight of the brain) and a measure of body size (e.g. body weight). Within a given taxonomic group of organisms, it is common to find

Size, Shape and Evolution

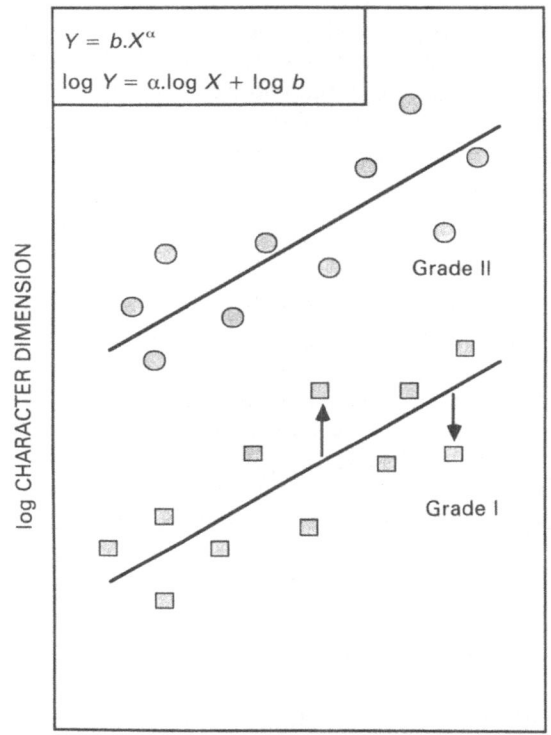

$Y = b.X^{\alpha}$

$\log Y = \alpha.\log X + \log b$

Grade II

Grade I

log CHARACTER DIMENSION

log BODY WEIGHT

Figure 8.1   Illustration of the basic principles of allometric analysis. For a given group of species (squares), a best-fit line has been determined for a bivariate data set plotted on logarithmic coordinates. The best-fit line indicates the idealised *scaling principle*, while positive or negative vertical deviations of individual species from the best-fit line (arrows) indicate special adaptations, reflected by the *residual values*. In many cases, when a second group of species is taken (circles), a best-fit line of similar slope is obtained, but it is vertically displaced with respect to the line determined for the first set. It is then appropriate to refer to the two groups of species as representing separate *allometric grades* (grade I; grade II). The vertical distance between the lines provides a measure of the magnitude of the *grade shift* involved.

that there is an obvious general trend in the magnitude of the character dimension with body size, but that there is also some scatter of the points for individual species. The general trend is represented by the best-fit line, while the special adaptations of individual species are indicated by their vertical distances above or below that line (namely by their *residual values*). Species lying above the best-fit line may be said to have greater than expected values of the character dimension for their body size, while species lying below the line may be said to have smaller than expected values. Using this simple approach, it may therefore be said that the best-fit line indicates the general scaling influence of body size, while the departures of individual species from the line (namely their residual values) indicate shifts in fundamental organisation. Hence, bivariate allometric analysis can in principle provide an empirical solution to the rhetorical question regarding size and fundamental biological organisation presented above.

Matters often become more complicated when a second group of species is introduced into the analysis. In some cases, it may be found that the same best-fit line will be applicable to the second group of species. Often, however, it is found that a different best-fit line is obtained. With remarkable regularity, it is found that the slopes of the best-fit lines are closely similar for the two groups of species, but that the lines are separated vertically. In other words, the value of the allometric exponent (α) is virtually the same in both cases, but the value of the allometric coefficient (b) is different. In such cases, the two sets of species can be described as representing distinct *allometric grades*. This difference, of course, is no more than a special case of the positive or negative departure of individual species from a best-fit line, but it is particularly revealing in that it coincides with taxonomic or other distinctions between groups of species. Here, it can be postulated that an allometric *grade shift* has occurred in the course of the evolution of the two groups of organisms concerned (Figure 8.1). It must be noted, however, that a bivariate plot of the kind shown in Figure 8.1 does not permit identification of the *direction* of evolution. It may be that grade II represents a shift away from the condition of grade I, or vice versa; it is also possible that the ancestral condition is no longer represented and that both grade I and grade II have shifted away from the original condition. Finally, it must always be remembered that different species may come to occupy the same allometric grade through convergent evolution.

It is necessary to provide some terminological clarification with respect to the value of the allometric exponent (α). Allometry actually refers to cases where this value differs from 1, while the special case of $\alpha = 1$ is referred to as *isometry* (Figure 8.2). Isometric scaling is straightforward proportional scaling, in which a simplified form of the allometric equation applies:

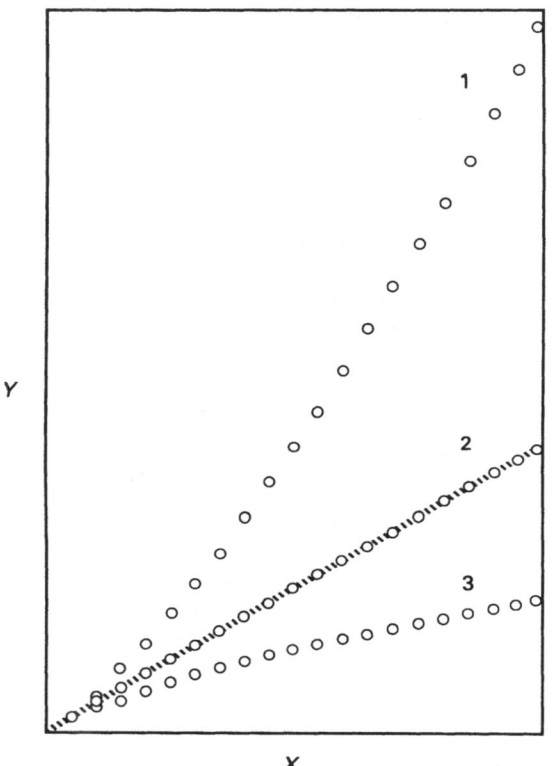

Figure 8.2   Curves illustrating the difference between positive allometry (curve 1; exponent = 1.3), isometry (line 2; exponent = 1) and negative allometry (curve 3; exponent = 0.75). Isometry is a special case in the family of allometric curves in that the exponent value is unity, such that a straight line relationship is obtained without logarithmic conversion. Ratios may therefore be used if an isometric relationship exists between two variables.

$$Y = b.X$$
$$\text{or } Y/X = b$$

If the allometric exponent is greater than 1 (corresponding to an accelerating curve), the relationship is said to be one of *positive allometry*; if the allometric exponent is less than 1 but greater than 0 (corresponding to a decelerating curve), the relationship is said to be one of *negative allometry* (Figure 8.2). There has been some confu-

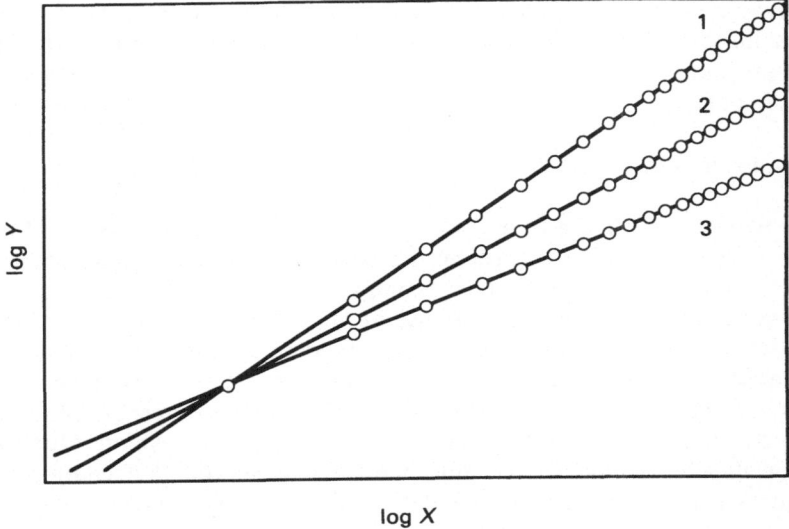

Figure 8.3 Logarithmic conversion of the curves in Figure 8.2 produces three straight lines. In each case, the slope of the line reflects the value of the allometric exponent.

sion in the literature because the term 'negative allometry' has sometimes also been used for cases in which the value of the allometric exponent is actually negative, representing an inverse relationship between Y and X. However, it is advisable to use the term 'inverse allometry' in such cases, especially because they usually arise through secondary manipulation of the data (see below). Huxley and Teissier (1936) in fact referred to an unusual example of intraspecific growth in crabs in which there is a decrease in the size of the abdominal limbs with increasing body size and proposed the term *enantiometry* (= negative growth) for this phenomenon. It seems likely, however, that this is a very rare occurrence essentially limited to the context of individual development and little would be gained by general use of this term for inverse allometric relationships.

Logarithmic conversion of the raw data plotted in Figure 8.2 generates the three straight lines shown in Figure 8.3. The slopes of the lines correspond to the allometric exponents in the relevant equations and the slope of the line for the isometric relationship is accordingly 1. This provides the basis for the standard test for isometry, in which it is established whether the slope of the line for a

Size, Shape and Evolution

bivariate logarithmic plot is significantly different from unity. If it is established that a given relationship is essentially isometric, a simple ratio (index) between Y and X can be used for comparative purposes (see below).

Ratios and indices

Having established a basic model for biological scaling, it is possible to turn to a number of problems associated with discussions of size relationships in evolutionary biology. One particular problem is associated with the widespread use of *ratios* and *indices*. As noted by Atchley *et al.* (1976), it is commonly implied, or even explicitly stated, that calculation of a ratio between two variables removes the effect of body size, such that values derived from animals of different body sizes can be directly compared in a meaningful fashion. Huxley specifically commented on this point and observed that caution is necessary. This is because calculation of ratios may not eliminate the effects of body size if there are underlying allometric relationships. Huxley cited the use of ratios in human craniometry and commented on a claim made by Parsons that evolutionary change had taken place in the skull of the British population during historic times. Although it is true that a progressive change can be observed in the cephalic index of the skulls concerned (namely in the ratio of breadth to length), it is also true that an accompanying increase in average body size can be documented in this population. Hence, if the components of the cephalic index (head breadth, head length) are allometrically related to body size, a change in the index may be an automatic consequence of increasing body size.

It is possible to recognise two different kinds of ratio that may be calculated. In the first case, biological parameter under investigation (Y) can be expressed as a simple ratio to body size itself (X):

$$Y = b.X^{\alpha}$$
$$\text{hence } Y/X = b.X^{\alpha-1}$$

It is easy to see that if the initial relationship between Y and X is allometric (namely $\alpha \neq 1$), the relationship between the derived ratio (Y/X) and X must also be allometric. Thus, the use of a ratio in this case simply converts one allometric relationship into another, more obscure, allometric relationship. The procedure is also statistically dubious, as the parameter X then appears on both sides of the equation and this necessarily biases any correlation that is determined. Ratios of this kind are commonly used in cases where a given parameter is expressed in terms of unit body weight. For instance, it is well established in comparative physiology that in mammals basal metabolic rate scales to body weight with an exponent value of approximately 0.75 (Benedict, 1938; Brody, 1945; Brody and Procter, 1932; Kleiber, 1932, 1947, 1961; Schmidt-Nielsen, 1984) – a rule which is commonly known as Kleiber's Law (see later). Because this is a negative allometric relationship ($\alpha < 1$), the implication is that, although basal energy needs increase progressively with increasing body size, the energy requirement per unit body weight decreases. This can be indicated directly by taking the 'mass-specific basal metabolic rate', which is the ratio between basal metabolic rate (M) and body weight (X):

$$M = b.X^{0.75} \quad \text{(Kleiber's Law)}$$
$$\text{hence } M/X = b.X^{-0.25}$$

The scaling of mass-specific basal metabolic rate provides an example of an inverse allometric relationship that is obtained by secondary manipulation of the data.

Alternatively, it is possible to define a ratio with two separate parameters (Y_1 and Y_2) each of which is related to overall body size (X). In this case:

$$Y_1 = b_1 X^{\alpha}$$
$$Y_2 = b_2 X^{\beta}$$
$$\text{hence } Y_1/Y_2 = (b_1/b_2)X^{\alpha-\beta}$$

It can be seen that, if α and ß take different values, the relationship between the ratio (Y_1/Y_2) and body size (X) must be allometric. The ratio will only be independent of body size if α and β have identical values. If Y_1 and Y_2 are independently related to body size in an allometric fashion, it is perfectly possible that they will scale in different ways. This must therefore be checked before a ratio is employed in the belief that it represents a 'size-free' index.

The question of the *statistical* properties of ratios was specifically addressed by Atchley *et al.* (1976). These authors concluded that the use of ratios can produce misleading results in a statistical sense, but Hills (1978) subsequently questioned some of their key arguments. In the light of this exchange of views about statistical properties of ratios, it is worth emphasising here that the use of ratios is questioned above on *biological*, not statistical, grounds. The simple fact of the matter is that it cannot be claimed that a ratio 'removes' the effect of body size if values of that ratio are found to vary systematically with body size.

In practice, it is very easy to determine whether a given ratio is free of residual body size effects or not. It is merely necessary to plot the empirical values obtained for the ratio against body size and to test whether any trend is still recognisable. Huxley (1932) in fact did this and showed that in British human skulls there is a clear tendency for values of the cephalic index to increase with overall head size. Thus, the observation that the cephalic index has tended to increase in the British population is no more than a trivial restatement of the fact that there has been a general tendency for skull size to increase, along with body size generally. Despite Huxley's warning, however, many subsequent investigators have used anthropometric ratios as if they represent size-free discriminators for meaningful discussion of differences within and between human populations.

A similar error has been made in many interspecific studies. A more complex example for such interspecific comparisons is provided by the *intermembral index*, which is the ratio of forelimb length (humerus + radius) to the hindlimb length (femur + tibia), expressed as a percentage. Values of this index have been commonly cited as indicators of locomotor adaptation of individual primate species, without any specific consideration of the potential effects of body size. However, it has been shown that there is a general trend for intermembral indices to increase with body size in primates generally (Rollinson and Martin, 1981). It is simply necessary to plot inter-membral indices against body weight on logarithmic coordinates to see that there is a general trend for values to increase with body size (Figure 8.4). In fact, the correlation coefficient for this plot is rela-tively high ($r = 0.61$) and the coefficient of determination ($r^2 = 0.38$) indicates that body size differences account for almost 40 per cent of the variation between species in intermembral index. This example corresponds to the second category of ratio indentified above (namely of the type Y_1/Y_2) and it is therefore clear that forelimb length (Y_1) must scale to body size in a different fashion from hindlimb length (Y_2). It also illustrates the point that scaling trends may not always be dictated by simple geometrical or developmental constraints. In this case, it is likely that the progressive increase in values of the intermembral index with increasing body size is attribut-able to some general selection pressure that favours relatively longer

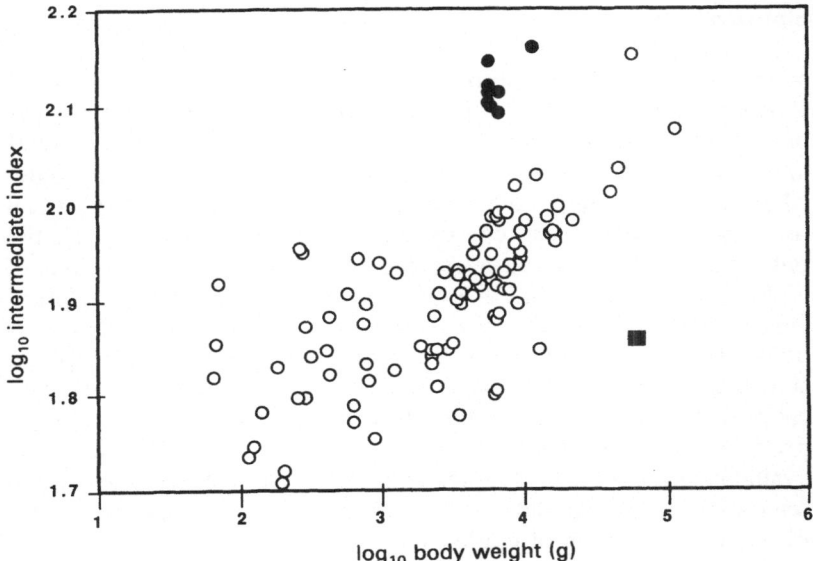

Figure 8.4 Bivariate logarithmic plot of intermembral indices against body weight for a sample of 103 living primate species. There is an obvious overall trend in the data, with the value of the intermembral index tending to increase with increasing body size ($r = 0.61$). The gibbons and siamang (black circles) have unusually high values for their body size, while *Homo sapiens* (black square) has an unusually low value relative to the body size.

forelimbs in larger-bodied species. Nevertheless, the general trend with body size should be both identified and taken into account in discussing the significance of intermembral index values for locomotor evolution in primates (Rollinson and Martin, 1981). In particular, it should be remembered that the significance of a particular value of the intermembral index will depend upon the body weight of the primate species concerned. For instance, the intermembral index values of gibbons (*Hylobates* spp.) are high and similar to those of great apes, but the gibbons are unusual in having such high values at a relatively small body weight. Further, the intermembral index of *Homo sapiens* is low and overlaps with the values determined for small-bodied, prosimian primates. Humans are, however, unique among primates in having such a low intermembral index combined with such a large body size.

Statistical problems

When it comes to allometric analysis itself, there are still major statistical problems that must be overcome. In fact, despite the burgeoning literature on the topic of scaling in primates and in other animals, the statistical aspects of allometric analysis have been somewhat neglected. Many authors make no mention at all of statistical problems and simply proceed to conduct allometric analysis as if the only real difficulties to be overcome were restricted to questions of biological interpretation. It is understandable that statistical problems were given scant attention at the time of publication of Huxley's book (1932), given the early stage of development of appropriate statistical techniques; but the widespread lack of attention to statistical problems in the modern literature is regrettable. The two major problems that are faced in any given study are: (1) Which of the available line-fitting techniques should be used for identifying the scaling trend in the data? (2) How can the investigator achieve objective recognition of grades in a given data set?

There is continuing discussion over the selection of an appriate best-fit line (e.g. see Harvey and Mace, 1982). If the data conform very closely to the best-fit line, the choice of line-fitting technique is largely irrelevant. However, if the data are appreciably scattered relative to the best-fit line, as indicated by comparatively low values of the correlation coefficient (r), marked divergence between the results obtained with different line-fitting techniques can occur. In practice, problems generally arise in cases where the value of the correlation coefficient (r) is less than 0.95. Most authors have used *regressions* for bivariate allometric analyses, as this is the best-known method and is relatively simple in its application. But the use of such an *asymmetrical* line-fitting approach for the analysis of biological parameters is open to question. Determination of a regression of Y on X is based on the twin assumptions that that Y is causally dependent upon X and that the independent variable X is measured without error. Neither condition is likely to apply in a case where two biological parameters are plotted against one another (e.g. age of sexual maturity or lifespan against body weight). In the first place, the causal relationships between two variables abstracted from a complex biological system are not usually known. Secondly, in most cases *both* parameters measured are often subject to exactly the same kind of measurement error (e.g. brain size and body size). It can therefore be argued that it is preferable to use a *symmetrical* line-fitting technique that does not presuppose an unlikely distinction between dependent and independent variables in a biological system and does not involve the spurious assumption that one variable is measured without error. The *major axis* is a symmetrically determined best-fit line that would seem to be generally appropriate for empirical determinations of interspecific scaling relationships.

Because of the persisting uncertainty surrounding the choice of a suitable line-fitting technique in bivariate allometric analysis, a valuable alternative approach is to fit a line of predetermined slope to the data (Martin and MacLarnon, 1988a, in press). Provided that some theoretical or empirical basis exists for inferring an expected value of the scaling exponent (α), it is possible to insert this fixed value into the allometric equation for a given data set and to determine the corresponding value of the allometric coefficient (b), given that the line must pass through the point corresponding to the mean values of the two biological parameters X and Y. In some cases, the expected exponent value may be indicated by geometrical considerations. For instance, scaling of a surface relative to a volume would normally be expected to entail an exponent value of 0.67. In other cases, it is possible that a widely recognised fundamental scaling relationship may be invoked to infer secondary scaling relationships. An example is provided by the scaling of basal metabolic rate (Kleiber's Law). Any allometric relationship involving a biological parameter that may reasonably be expected to depend upon basal metabolism might be expected to show the same scaling exponent (0.75) as that for the relationship between basal metabolism and body weight.

A line of predetermined slope may also be used in a case where two or more grades are apparently present in a given data set. Use of a single line with an appropriate slope in such cases permits direct comparison of all points in the complete data set. In cases where the empirical best-fit lines determined for the individual grades within the data set do not have exactly the same slopes, such a procedure is particularly helpful. The slope of the single line fitted to the complete data set can either be taken as the average of the values determined for the individual grades or be predetermined on the basis of some theoretical expectation.

This brings us to the second major problem encountered in allometric analysis, in that a given data set may be composed of two or more subsets. As has been noted above, it is often found that separate lines of approximately the same slope may be fitted to data derived from different taxonomic groups of animals. In such cases, it may be said that each group obeys the same general scaling principle (as is indicated by closely similar values for the allometric exponent, α), but the taxonomic groups differ from one another in the specific nature of the scaling relationship (as reflected by the value of the allometric coefficient, b). The need to define a widely recognised, objective technique for identification of separate grades of this kind in a given data set is one of the main priorities for the future development of allometric analysis. A single best-fit line determined for an entire data set that in fact includes two or more distinct subsets (grades) will usually have an inappropriate slope value. Although the overall slope value will, of course, reflect a general trend in the data, it will fail to reflect the scaling principle operating within each grade containing functionally comparable animals. Thus, there is a potential problem of 'grade confusion' in any allometric analysis. Incorrect identification of the scaling principle as a result of grade confusion will of course lead on to the determination of misleading residual values for individual species. A similar problem exists with the presence of extreme outliers, which will necessarily bias the slope of any best-fit line that is determined.

In conclusion, therefore, the successful application of allometric analysis to any bivariate data set requires not only recognition of the problems attached to selection of an appropriate line-fitting technique but also awareness of possible distortion of the results through effects of grades and outliers.

APPLICATIONS OF ALLOMETRIC ANALYSIS

The technique of allometric analysis, if appropriately applied, permits the investigator to take the scaling influence of body size into account and to achieve more informative comparisons between species. The author, in concert with various collaborators, has now applied this technique to a wide range of problems in primate biology including morphology of the skull, brain size, dimensions of the gastrointestinal tract, general reproductive biology and fetal development (e.g. see Harvey et al., 1987; MacLarnon et al., 1986a, 1986b; Martin, 1980, 1981, 1982, 1983, 1984a, 1984b; Martin and Harvey, 1985; Martin and MacLarnon, 1985, 1986, 1988a, 1988b, Martin et al., 1985; Ross, 1988; Rudder, 1979). From these and other studies, it has emerged that a series of practical applications can be identified, as follows:

(1) *Recognition of general scaling principles.* The scaling relationship for a given data set is characterized primarily by the empirically determined value of the allometric exponent (α). If a particular kind of scaling relationship emerges repeatedly from a number of different studies, and particularly if there is some theoretical justification for a given exponent value close to that indicated empirically, it is reasonable to conclude that some general scaling principle may be involved. It must be noted, however, that recognition of general scaling principles requires special attention to the possible presence of grade effects (see below).

(2) *Recognition of outliers.* The distances of individual points from the best-fit line (namely their residual values) provide information about the special adaptations of individual species relative to the general trend. Extreme cases of points lying far from the line (outliers) are of particular interest and demand biological interpretation.

(3) *Recognition of grades.* In any given data set, it is necessary to test for the presence of grades. This can be done by dividing the data set into various sub-sets (e.g. on taxonomic grounds) and seeking differences between them. In some cases, it may also be possible to apply objective procedures for revealing the presence of grades (see below).

(4) *Testing of hypotheses.* It is possible to conduct allometric analysis in such a way as to test preformulated hypotheses, which may concern the scaling principle, the occurrence of particular species as outliers, or the existence of grade distinctions between different groups of species.

(5) *Prediction of unknown values.* Given an established allometric relationship between a given parameter and body size, it is possible to use the empirically determined best-fit line to derive a prediction of the expected value of the parameter for a species for which only the body size is known. Alternatively, if body size itself is unknown (e.g. for fossil species) the value of the parameter can be used to obtain an estimate of body size. In both cases, however, it must be remembered that the best-fit line only describes a general trend and that the degree of scatter around the line will be matched by a comparable degree of uncertainty concerning the predicted value.

(6) *Inference of functional relationships.* It is, in principle, possible to infer functional relationships on the basis of empirically established scaling relationships. But it must never be forgotten that

bivariate analyses are concerned only with *correlation* and not directly with *causal relationships*. It is easy to misread the direction of a correlation and hence to draw erroneous conclusions. Indeed, it is perfectly possible for two variables subjected to allometric analysis (Y_1, Y_2) to be causally linked through some third variable (X) that has not been examined. Causal relationships must therefore be inferred with extreme caution and they should ideally be confirmed by a series of allometric analyses (rather than by a single bivariate study) and refined by repeated testing. It should also be noted that uncertainty about the causal relationship between two biological variables in fact conflicts with one of the basic requirements for the use of linear regression, as noted above.

These different applications of allometric analysis can be illustrated with a series of practical examples.

Recognition of a scaling principle

General scaling principles may be based on a priori theoretical considerations, as is the case (for example) with geometrical scaling. In some cases, however, a scaling principle may be recognised in the absence of explicit theoretical underpinning, on the grounds that empirical studies have repeatedly indicated the presence of a particular kind of relationship. Perhaps the best known example of such an empirically determined general scaling principle has already been mentioned above, namely Kleiber's Law governing the scaling of basal metabolic rate (M) to body weight (W) in mammals:

$$M = b.W^{0.75}$$

Kleiber in fact proposed the following standard relationship for placental mammals:

$$M = 70.W^{0.75}$$

(where W is measured in kg and M in kcal/day)
    This formula converts to the following equation:

$$M = 3.42.W^{0.75}$$

(where W is measured in g and M in ml O_2/h)

Kleiber's Law, with its exponent value of approximately 0.75, is widely accepted and applied in comparative physiology (e.g. Benedict, 1938; Brody, 1945; Brody and Procter, 1932; Kleiber, 1932, 1947, 1961; McNab, 1980, 1983, 1986; Schmidt-Nielsen, 1984). Indeed, it is common to refer to 'metabolic body weight' ($W^{0.75}$) rather than actual body weight in studies dealing with metabolic requirements and turnover, as it is reasonable to expect that a large variety of secondary parameters will be directly proportional to basal metabolic rate. Although Kleiber's Law is widely accepted and applied, however, there is no generally acknowledged theoretical explanation for the particular value of the exponent (taken, for ease of calculation, to be 3/4 or 0.75). Nevertheless, in a recent study, Elgar and Harvey (1987) have once again confirmed with a large sample of 265 species from 18 orders of mammals that the scaling of basal metabolic rate is negatively allometric and that the exponent value is close to 0.75. For species within genera, for generic averages within families and for familial averages within orders, they determined scaling exponents of 0.73, 0.75 and 0.73, respectively. For overall average values for the 18 mammalian orders, a somewhat higher scaling exponent value of 0.83 was obtained, but the 95 per cent confidence limits on this value (0.76–0.92) would permit a value just in excess of the conventional figure of 0.75 generally accepted for Kleiber's Law.

The scaling of basal metabolic rate to body weight is illustrated for 25 primate species and 3 tree-shrew species in Figure 8.5 (data derived from: Daniels, 1984; Dobler, 1982; Kurland and Pearson, 1986; McCormick, 1981; Müller, 1983, 1985; Müller *et al.*, 1983; Müller *et al.*, 1985; Nagy and Milton, 1979; Richard and Nicoll, 1987). It can be seen that this sample conforms quite closely to the expectation from Kleiber's equation, though there are some deviations from the line. A line of fixed slope = 0.75 fitted to the data for the 25 primate species yields the formula:

$$\log_{10}M = 0.75.\log_{10}W + 0.420$$

equivalent to:

$$M = 2.63.W^{0.75}$$

Hence, in comparison to the standard formula for Kleiber's Law given above, it can be seen that for primates overall basal metabolic rates are approximately 77 per cent of the values predicted from Kleiber's Law.

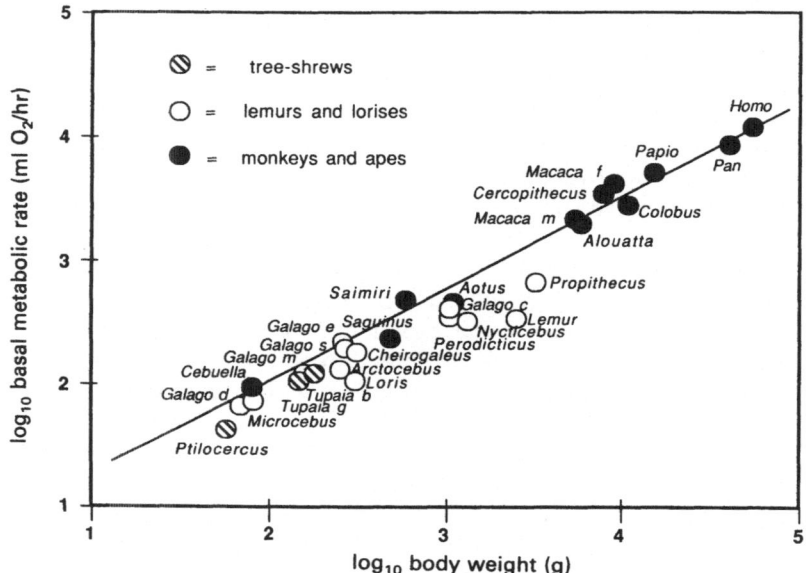

Figure 8.5   Bivariate logarithmic plot of basal metabolic rate against body weight for a sample of 25 primate species and three tree-shrews. The line represents the expectation from the Kleiber equation for placental mammals. It can be seen that the simian primates (monkeys and apes) conform quite closely to expectation, whereas the strepsirhine primates (lemurs and lorises) generally tend to lie below the line.

Recognition of individual outliers

Kleiber's Law was based on a series of placental mammals from the northern hemisphere, several of them being domestic forms, and these happened to conform quite closely to a single best-fit line (see Kleiber, 1961). It has since emerged, however, that there are numerous placental mammals that diverge quite markedly from Kleiber's line, usually lying below it. This is the case, for instance, with the specialised leaf-eating sloths (McNab, 1978), which have basal metabolic rates less than 50 per cent of the level predicted by the Kleiber equation. McNab (1986), supported by Kurland and Pearson (1986), has suggested as a general explanatory principle that relatively low basal metabolic rates are linked to specific dietary habits, such as leaf-eating (folivory); but Elgar and Harvey (1987) have concluded from a subsequent re-analysis of a large data set that variations in

1. T. H. Huxley with his grandson, Julian, in 1895 (from L. Huxley, *Life and Letters of Thomas Henry Huxley*, London, 1900)

2. Aldous and Julian Huxley in 1958 (*photograph by Douglas Glass*)

basal metabolic rate, relative to body size, largely reflect taxonomic affinities. There is also considerable variation among primates in basal metabolic rate relative to body size, as can be seen from a list of residual values determined relative to the Kleiber equation (Table 8.1). Strepsirhine primates of the subfamily Lorisinae (*Arctocebus*, *Loris*, *Nycticebus*, *Perodicticus*) are notable for their relatively low basal metabolic rates, and among simian primates the owl monkey (*Aotus*) has an unusually low basal metabolic rate. Old World monkeys, by contrast, generally tend to have relatively high basal metabolic rates.

Recognition of grades

With basal metabolic rate, it is possible to recognise grades as well as individual outliers among mammals. For primates, Müller (1983, 1985) has shown that the strepsirhines (lemurs and lorises) generally have lower basal metabolic rates than simians (monkeys, apes and man), and this is readily apparent from Table 8.1 and from Figure 8.5. In simians, basal metabolic rates conform fairly closely to the level expected from Kleiber's Law, in some cases even exceeding this level, while in strepsirhine primates the values generally lie between 20 per cent and 60 per cent below expectation. A line of fixed slope = 0.75 fitted to the data for the 12 simian primate species included in Figure 8.5 has the following formula:

$$\log_{10}M = 0.75.\log_{10}W + 0.520$$

equivalent to:

$$M = 3.31.W^{0.75}$$

In other words, values for simian primates overall tend to be about 3 per cent lower than those predicted by Kleiber's Law. By contrast, a line of fixed slope = 0.75 fitted to the data for the 13 strepsirhine primate species included in Figure 8.5 has the following formula:

$$\log_{10}M = 0.75.\log_{10}W + 0.334$$

equivalent to:

$$M = 2.16.W^{0.75}$$

Table 8.1	Basal metabolic rates of primates and tree-shrews expressed
as percentages of the values expected from the Kleiber equation
(see Figure 8.5)

Species	Taxonomic group	BMR as percentage of Kleiber
Lemur fulvus	Lemur	28.3
Nycticebus coucang	Loris	42.2
Loris tardigradus	Loris	45.1
Propithecus verreauxi	Lemur	47.0
Ptilocercus lowii	Tree-shrew	60.4
Perodicticus potto	Loris	61.8
Saguinus geoffroyi	New World monkey	66.5
Arctocebus calabarensis	Loris	70.5
Galago crassicaudatus	Loris	70.9
Tupaia belangeri	Tree-shrew	73.1
Aotus trivirgatus	New World monkey	75.4
Tupaia glis	Tree-shrew	77.2
Galago demidovii	Loris	78.0
Microcebus murinus	Lemur	78.4
Cheirogaleus medius	Lemur	79.2
Galago moholi	Loris	79.6
Colobus guereza	Old World monkey	84.3
Tarsius spectrum	Tarsier	85.6
Galago senegalensis	Loris	85.8
Alouatta palliata	New World monkey	87.8
Pan troglodytes	Old World ape	89.1
Galago elegantulus	Loris	97.7
Homo sapiens	Human	101.5
Macaca mulatta	Old World monkey	102.4
Cebuella pygmaea	New World monkey	103.0
Cercopithecus mitis	Old World monkey	112.1
Papio ursinus	Old World monkey	112.4
Saimiri sciureus	New World monkey	119.7
Macaca fuscata	Old World monkey	131.2

In other words, values for strepsirhine primates overall tend to be
only about 63 per cent of those predicted by Kleiber's Law and
approximately 65 per cent of those for simian primates generally.
Although there is some scatter among the strepsirhine primates
(Figure 8.5), it is clear that simians overall have higher basal metabolic rates and conform more closely to the expectation from the
Kleiber equation. Müller (1985) considered two main explanations
for this difference between simian and strepsirhine primates, the first

being that the strepsirhine primates possess a more primitive type of bodily organisation and the second being that they are in fact special-ised to meet specific ecological requirements. He concluded that the low basal metabolic rates of strepsirhine primates – along with other features such as less precise regulation of body temperature and general lethargy – represent a special adaptation rather than a primi-tive condition. Further, in agreement with McNab, Müller concluded that lowering of basal metabolic rate as a special adaptation occurs in response to specific ecological parameters such as dietary resources. Nevertheless, it should be pointed out that the difference between strepsirhine primates and simian primates corresponds to a tax-onomic distinction, such that arguments implying separate adapta-tion of all the individual species compared may not be justifiable (see Elgar and Harvey, 1987).

It should also be noted that all three tree-shrew species included in Figure 8.5 lie below the Kleiber line. On average, tree-shrews exhibit basal metabolic rates that are only about 70 per cent of the values predicted by the Kleiber equation. On the one hand, this provides yet another example of relatively low basal metabolic rates characteris-ing a particular taxonomic group. On the other, it suggests that relatively low basal metabolic rates (compared to the Kleiber equa-tion) might conceivably be a primitive feature of placental mammals generally, as tree-shrews exhibit numerous primitive mammalian features and may be presumed to have retained a relatively primitive lifestyle.

A particularly clear case of grade distinction is provided by the scaling of mammalian gestation periods (Martin and MacLarnon, 1985). It has been general practice in the literature on the scaling of reproductive parameters in placental mammals to cite an overall relationship between gestation period and body weight, with a scaling exponent value of approximately 0.25 (Martin, 1981; Martin and MacLarnon, 1986, in press). This is illustrated in Figure 8.6 (upper graph). Determination of a single best-fit line, however, ignores the fundamental distinction between altricial and precocial mammals (Portmann, 1938, 1939, 1941, 1965). Altricial neonates are born in a poor state of physical development, typically in multiple litters, while precocial neonates are born in a relatively advanced state of develop-ment, typically as singletons (Martin, 1983; Martin and MacLarnon, 1985). It is therefore to be expected that gestation periods for altricial mammals should be shorter, relative to body size, than those for precocial mammals. As was noted by Portmann, a particular type of

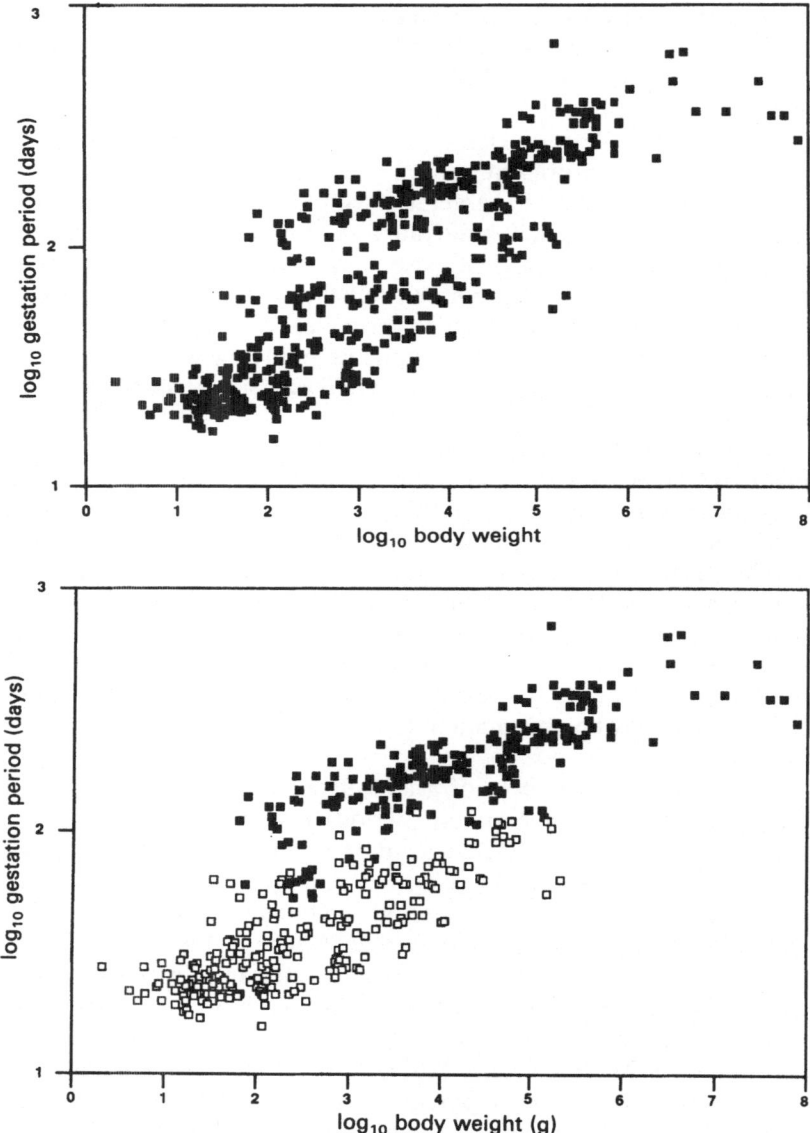

Figure 8.6   Plots of gestation period against body weight for a sample of
429 species of placental mammals. In the upper graph, the sample is taken
to represent a single distribution. A single best-fit line determined for this
plot has a slope of 0.22. In the lower graph, the sample has been divided

neonate generally characterises large taxonomic groups (orders or suborders) of mammals. Carnivores, insectivores, tree-shrews, lagomorphs, myomorph rodents and sciuromorph rodents are generally characterised by altricial offspring, whereas cetaceans, hoofed mammals, primates, hystricomorph rodents, pinnipeds, elephants and manatees are generally characterised by precocial offspring. When the sample of mammalian gestation periods shown in the upper graph of Figure 8.6 is divided into precocial and altricial species on the basis of these taxonomic categories, it becomes clear that there are two grades present in the data (Figure 8.6, lower graph). Indeed, the difference between the two grades is so pronounced that a typical precocial mammal is found to have a gestation period approximately four times longer than a typical altricial mammal of the same body weight.

In the case of gestation periods, it is possible to apply an objective technique for the recognition of grade distinctions (Martin and Mac-Larnon, 1985). This technique is based upon the fact that all best-fit lines must pass through the point representing the mean values for the two parameters X and Y. Using an iterative computer routine, it is possible to fit a line of fixed slope passing through this fixed point and to vary the slope of the best-fit line in a step-wise fashion. For each slope value, the residuals of all species are calculated and presented in histogram form. If the set of points constitutes a single distribution, all histograms obtained in this way should remain unimodal. If, on the other hand, the set of points contains two clearly separated subsets of points, the pattern of residuals should become biomodal at the slope value representing the scaling principle for the separate subsets. When this technique is applied to the data for mammalian gestation periods (Figure 8.7), a unimodal distribution of residuals is obtained with a slope value of 0.25, but a clearly bimodal histogram is obtained with slope values of 0.15 and 0.10. As the latter value is close to the average value obtained for the scaling of gestation period within individual orders and suborders of mammals

---

between taxonomic groups characterised by altricial offspring (white squares: carnivores, insectivores, tree-shrews, lagomorphs, myomorph rodents, sciuromorph rodents; $N = 227$) and those characterised by precocial offspring (black squares: cetaceans, hoofed mammals, primates, hystricomorph rodents, pinnipeds, elephants and manatees; $N = 222$). It can be seen that at least two grades are present in the data, each with a markedly lower slope than that for an overall best-fit line determined for the complete data set. (Adapted from Martin and MacLarnon, in press.)

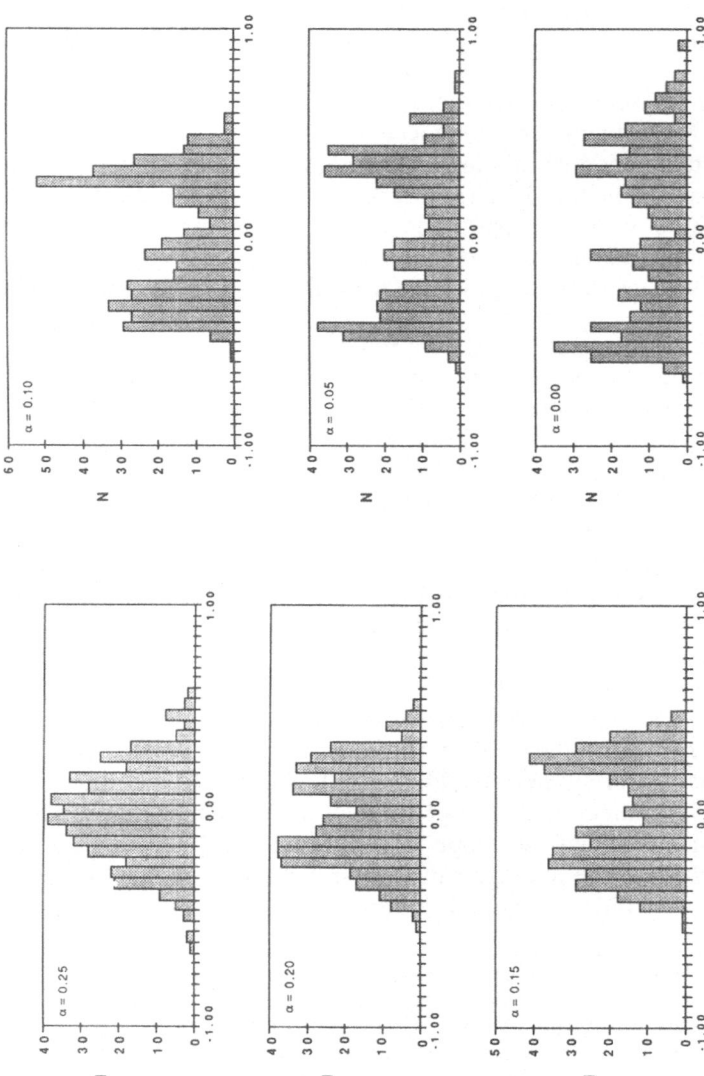

Figure 8.7 Histograms of residual values for mammalian gestation periods ($N = 427$) calculated relative to lines of fixed slope ranging from 0.25 down to 0.00, by decrements of 0.05. With a slope value of 0.25, the distribution of residual values is unimodal. As the slope value decreases, the distribution becomes bimodal, with the deepest trough between the two halves of the distribution at a slope value of approximately 0.10. At low slope values (0.05, zero), the histogram is at least trimodal and possibly quadrimodal. The histogram with a slope value of zero is equivalent to the distribution of logarithmic values of gestation periods determined without taking body size into account. (See text for discussion.)

characterised by a particular neonate type, the following general scaling relationship between gestation period (G) and body weight (W) can be proposed:

$$G = b.W^{0.10}$$

It should also be noted from Figure 8.7, that a histogram of mammalian gestation periods generated without taking the effects of body size into account (namely with a scaling exponent of zero) is actually trimodal or quadrimodal. The first peak represents small-bodied altricial mammals and the third peak represents precocial mammals of medium body size. However, overlap between large-bodied altricial mammals and small-bodied altricial mammals generates an intervening second peak (Martin and Maclarnon, in press), while very large-bodied precocial mammals are to some extent isolated and generate a small fourth peak.

Testing of hypotheses

As has been indicated above, the allometric model can be applied in a number of different ways to test hypotheses. The model can be used to test a hypothesis regarding a particular scaling principle (*exponent test*), or it may be used to test hypotheses regarding the departure of a single point (outlier) or set of points (grade) from a given reference line (*coefficient test*).

An example of a hypothesis predicting a particular exponent value is provided by the relationship between neonatal weight (W_N) and gestation period (G). Let it be assumed that neonatal weight can be taken as an approximate measure of the total resources devoted to the growth of the individual fetus by the mother during gestation. Let it also be assumed (following Payne and Wheeler, 1967) that the resources devoted a fetal growth can be averaged over the gestation period to yield an indication of maternal investment per unit time (W_N/G). One might reasonably expect daily investment in fetal growth to be related to the mother's overall metabolic capacity. Although this expectation actually relates to the mother's *total* metabolic turnover, it is conceivable that basal metabolic rate (M) would provide an approximate indication of her total metabolic capacity. Indeed, because a mother must support her developing fetus at all times, including when she is inactive (i.e. exhibiting a basal level of metabolism), it can be argued that basal metabolic rate would actually be a more sensitive indicator of her capacity to invest in fetal growth (Martin and MacLarnon, 1988a). On the basis of these considerations, it is possible to predict the following relationships:

daily maternal investment $(W_N/G) \propto M$

given that: $M \propto b.W^{0.75}$ (Kleiber's Law)

it follows that: $W_N/G \propto W^{0.75}$

Accordingly, it is to be expected that a best-fit line determined for the relationship between W_N/G and maternal body weight should have an exponent value of approximately 0.75. When this expectation is tested using 26 of the 28 mammal species in the sample listed in Table 8.1 (relevant reproductive data being lacking for *Galago elegantulus* and *Ptilocercus lowii*), the following empirical scaling relationship is found, with a relatively strong correlation ($r = 0.96$):

$$W_N/G = 2.6.W^{0.76} \times 10^{-3}$$

The hypothesis is thus supported by the empirical analysis.

In fact, this particular example also provides illustrations of predictions regarding grades and outliers. It has already been demonstrated above that strepsirhine primates (lemurs and lorises) generally have lower basal metabolic rates, relative to body weight, than simian primates (monkeys, apes and man). Hence, it is to be expected that strepsirhine primates might be generally constrained to exhibit lower daily maternal investment in gestation, relative to body weight, than simian primates. It has indeed been known for some time (Leutenegger, 1973; see also Martin and MacLarnon, 1988a, in press) that, relative to maternal body weight, strepsirhine primates produce markedly smaller neonates than simians. Nevertheless, because there is considerable scatter in the relationship between gestation period and maternal body weight – especially among strepsirhine primates – it is necessary to examine the relationship between the measure W_N/G and maternal body weight for strepsirhines and simians in order to obtain a clearer picture of daily maternal investment in fetal growth. This can be done by examining the ratio between W_N/G and body size adjusted for the expected scaling of basal metabolic rate ($W^{0.75}$). Given that the prediction outlined above requires that there should be an overall isometric relationship between these two var-

Table 8.2   Ratios of daily maternal investment (W_N/G) to 'metabolic body weight' ($W^{0.75}$) for primates and tree-shrews

Species	$W_N/G \times 100$	$W^{0.75}$	Ratio
Loris tardigradus	6.6	64.0	0.103
Perodicticus potto	24.1	173.9	0.139
Nycticebus coucang	25.8	152.1	0.170
Lemur fulvus	62.8	344.9	0.182
Propithecus verreauxi	76.4	412.9	0.185
Galago crassicaudatus	36.7	173.8	0.211
Galago moholi	9.3	44.1	0.211
Colobus guereza	197.7	854.0	0.232
Galago senegalensis	16.4	67.5	0.243
Saguinus geoffroyi	26.2	103.0	0.254
Pan troglodytes	742.9	2844.3	0.261
Arctobecus calabarensis	18.8	63.6	0.295
Galago demidovii	6.8	22.3	0.305
Macaca fuscata	286.7	931.7	0.308
Papio ursinus	456.7	1340.0	0.341
Homo sapiens	1264.0	3591.5	0.352
Alouatta palliata	258.1	666.7	0.387
Cebuella pygmaea	10.9	26.5	0.411
Aotus trivirgatus	72.9	176.6	0.413
Microcebus murinus	10.8	24.7	0.437
Macaca mulatta	287.5	633.9	0.454
Tupaia belangeri	22.7	49.6	0.458
Cercopithecus mitis	243.6	529.2	0.460
Tupaia glis	20.0	41.8	0.479
Saimiri sciureus	65.7	119.7	0.549
Cheirogaleus medius	30.9	47.7	0.648

iables, it is justifiable to use a simple ratio in this case. The expectation is that species with relatively low basal metabolic rates should also have a low value of the ratio. Ratios calculated in this way are listed in ascending order in Table 8.2. It can be seen, in comparison with Table 8.1, that primate species with relatively low basal metabolic rates also tend to be low-ranking with respect to the ratio between W_N/G and $W^{0.75}$. There is a general tendency for strepsirhine primates to rank lower than simian primates overall, as predicted. This result is supported by a more detailed study conducted by Rasmussen and Izard (1988) on the scaling of growth and life history traits in primates belonging to the family Lorisidae. The lorises studied (*Loris* and *Nycticebus*) are characterised by particularly low basal metabolic rates in comparison to the bushbabies studied (*Ga-*

lago crassicaudatus and *Galago senegalensis*) and it was found that these lower metabolic rates were correlated with slower development both prenatally and postnatally.

Nevertheless, some strepsirhine primates (notably *Cheirogaleus medius* and *Microcebus murinus*) do exhibit quite high values for the ratio between maternal investment in fetal growth and metabolic body size. (Table 8.2). These two species do not rank highly with respect to basal metabolic rate, expressed as a percentage of the figure expected from the Kleiber equation (Table 8.1); so they do not conform to expectation. It can therefore be concluded that, although strepsirhine primates generally conform to the expectation that relatively low basal metabolic rates will be reflected in relatively low daily investment in fetal growth, there are some exceptions which require explanation. Similarly, the two tree-shrew species examined (*Tupaia belangeri* and *T. glis*) rank quite highly with respect to the ratio between W_N/G and $W^{0.75}$ (Table 8.2), but are low-ranking with respect to basal metabolic rate, expressed as a percentage of the figure expected from the Kleiber equation. Once again, the relative rankings do not fit the prediction and it is clear that some other factor must be involved (see later).

The relationship between neonatal body weight and gestation period among primates is also of interest with respect to at least one conspicuous outlier – the tarsier (*Tarsius*). The gestation period of *Tarsius* has for some time been cited as approximately six months, but only on the basis of very limited, indirect data. A gestation period of six months would at first sight appear to be unusually long for such a small-bodied primate (approx. 140 g). The general scaling relationship between gestation period and body weight for 31 simian primate species (Hunter *et al.*, 1979) would indicate a gestation period of only 120 days for a species with the body weight of *Tarsius*. Recently, however, detailed studies of the reproduction of *Tarsius bancanus* in captivity have yielded a reliable gestation period of 178 days (Izard *et al.*, 1985). The average neonatal body weight for the genus *Tarsius* is approximately 28 g and coincides quite closely with the expectation from the general scaling relationship in relation to maternal body weight for simian primates (Leutenegger, 1973). If the gestation period were, indeed, 120 days as expected from the body weight of *Tarsius* in comparison to the general relationship for simian primates, the ratio of neonatal weight to gestation period ($W_N/G \times 100$) would be 20.8. If this measure of expected daily maternal investment in fetal growth ($W_N/G \times 100$) were to be expressed as a ratio to 'metabolic body size' ($W^{0.75}$), a value of 0.511 would be obtained, implying relatively heavy maternal investment. As can be seen from Table 8.2, this would be one of the highest values recorded for primates.

In fact, however, it seems very likely that tarsiers have relatively low basal metabolic rates. Clarke (1943) reported preliminary results indicating that *Tarsius spectrum* may have a basal metabolic rate lower than expected from the Kleiber equation. Even though the animals were probably kept at a temperature below that of the thermoneutral zone, and were therefore probably using extra energy to maintain the body temperature, Clarke's figures indicate that the metabolic rate was at least 15 per cent below the basal level predicted from the Kleiber equation (Table 8.1). Recent observations of the food intake of *Tarsius bancanus* (Roberts, 1988) have now indicated that basal metabolic requirements in *Tarsius* may be markedly lower than expected for a mammal of this body size. Thus, although the basal metabolic rate of *Tarsius* has yet to be measured under standardised conditions, it would seem very likely that tarsiers have relatively low basal metabolic rates and that there is therefore a constraint on maternal investment in maternal growth. It is probably for this reason that *Tarsius* has a gestation period that is approximately 50 per cent longer than expected (178 days, as opposed to 120 days). This relative extension of the gestation period brings the ratio between daily investment in fetal growth (W_N/G) and 'metabolic body size' ($W^{0.75}$) down to 0.382, which is an intermediate level in comparison to other primates. In this way, *Tarsius* is able to produce a neonate that is approximately of the size expected in comparison to simian primates.

An alternative approach to hypothesis-testing is to examine the relationship between the logarithmic residual values for two variables (Y_1, Y_2), having removed the effect of body weight (X) from both using bivariate allometric analysis. It may, for instance, be predicted that species with positive residual values for variable Y_1 will also tend to have positive residual values for variable Y_2 (Figure 8.8). This prediction can be tested by establishing the degree of correlation between the logarithmic residual values for Y_1 and Y_2. This is equivalent to calculating a partial correlation between Y_1 and Y_2, having removed the effect of X, using logarithmic values throughout. (I am indebted to Dr P. H. Harvey for clarifying this correspondence between the allometric procedure of correlation of residuals and the standard statistical procedure of partial correlation.) Taking the example of daily maternal investment in fetal growth (W_N/G) in

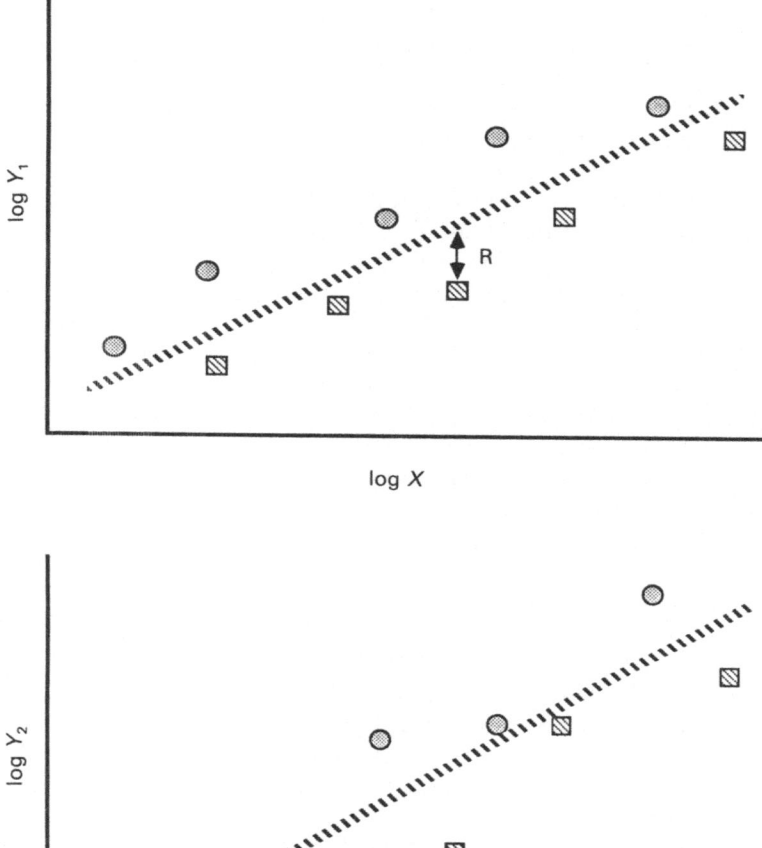

Figure 8.8 Illustration of the procedure of correlation of allometric residuals. Using separate bivariate analyses (upper graph; lower graph), it is possible to take account of the effects of body size (X) on two variables (Y_1 and Y_2) by calculating residual values (R). If species that tend to lie above the line for variable Y_1 also tend to lie above the line for variable Y_2 (circles *versus* squares in both plots), the residuals for the two variables will be strongly correlated. (See text for further discussion.)

relation to basal metabolic rate in primates, it is possible to examine the relationship between residual values for both variables, having first removed the effect of maternal body weight. In fact, as both variables can predicted to scale to maternal body weight with an exponent value of approximately 0.75, residual values can be determined relative to this theoretical expectation, using a line of fixed slope. When this was done for 18 primate species (Martin and MacLarnon, 1988), it was found that the residual values of daily maternal investment and basal metabolic rate are quite highly correlated ($r = 0.72$; $p < 0.001$). This indicates that variation in basal metabolic rate, relative to maternal body weight, can account for about 50 per cent of the observed variation in the measure W_N/G, following correction for the effect of body size.

An alternative, neatly illustrative, example of the testing of hypotheses through examination of the correlation between residual values is provided by Harvey and Zammuto (1985). These authors set out to test the hypothesis that age at first breeding is positively related to lifespan in natural populations of mammals. Faced with the problem that age at first breeding and life span both vary allometrically with body weight, Harvey and Zammuto first scaled the two variables to body weight and then examined the relationship between their residual values. They found that the two variables were, indeed, still strongly correlated after taking account of body size effects and thus seemed to confirm their initial hypothesis. It was subsequently pointed out (Sutherland *et al.*, 1986) that there is a statistical problem in this analysis in that total lifespan includes the time taken to reach sexual maturity, such that a spurious correlation between the residuals for these two measures might arise. Testing of the residuals of reproductive lifespan (i.e. total lifespan minus age at first breeding) against the residuals of age at first breeding yielded a reduced, but none the less highly significant, correlation between them.

Prediction of unknown values

Having established a general scaling relationship for a particular variable, it is possible to make predictions for species not included in the original sample. For instance, Hunter *et al.* (1979) drew upon an empirical scaling relationship for gestation period in relation to body weight, established for 31 simian primate species, to predict the (previously unknown) gestation period of the owl monkey (*Aotus trivirgatus*). The predicted value of 150 days was reasonably close to the first accurately timed gestation period of 133 days. In fact, it would seem that the average gestation period of *Aotus* is closer to 130 days than 150 days, and this indicates that the gestation of the owl monkey is somewhat shorter than expected for a monkey of that body size.

The prediction of unknown values has been particularly prevalent in the inference of body weight for fossil primate species, using empirical scaling relationships established for living primates between body weight and dental, cranial or skeletal parameters (e.g. Aiello, 1981; Conroy, 1987; Gingerich, 1977; Gingerich *et al.*, 1982; Gingerich and Smith, 1985; Martin, 1980, 1982). In this case the scaling relationship is the inverse of that usually applying in allometric analysis; body weight is the *Y*-variable and the dental, cranial or skeletal measure used is the *X*-variable. This approach has been used with considerable success in the inference of approximate body weights for fossil primate species, but it has its pitfalls and may sometimes yield misleading estimates. (See Smith, 1985, for an illuminating discussion of the problems.) Particular care must be taken, for example, in the inference of body size from dental measures, as there are numerous grades and outliers in the scaling relationships between dental parameters and body size. This results in lower correlations between dental measures and body size than commonly found with cranial or skeletal measures (Martin, 1980, 1982). It is known, for instance, that tarsiers have conspicuously large molar teeth for their body size (Gingerich *et al.*, 1982; Gingerich and Smith, 1985), such that inference of body size on the basis of molar dimensions would lead to marked overestimation. Accordingly, it is best to check estimations of body weights for fossil species in a number of different ways (Martin, 1980), rather than relying on inference from a single dental variable.

Inference of functional relationships

As has been pointed out above, the inference of functional relationships represents the most problematic application of allometric analysis, because there is a considerable gulf between the demonstration of a *correlation* between two variables and the establishment of a *causal relationship* between them. The utmost caution must be exercised in proposing causal relationships on the basis of interspecific comparisons, rather than through direct experimental investigation of causal connections in single species. The primary advantage of interspecific analysis resides not in the demonstration of causality but in the recognition of overall associations that cannot easily be recognised (if at all) from studies of single species (e.g. see Martin, 1983).

It is nevertheless possible to establish working hypotheses regarding functional relationships, on the basis of results from an initial allometric analysis, and to proceed to test these hypotheses by means of further comparative analyses. Although it remains exceedingly difficult to demonstrate causal connections, it is possible to narrow down the options. A case in point is provided by the scaling of brain size to body size in placental mammals. It was until recently widely accepted that the scaling exponent for the allometric relationship between mammalian brain size and body size is close to 0.67 (e.g. see Jerison, 1973), suggesting some kind of link between brain size and surface areas of the body. Eventually, however, it was shown with detailed analyses of large samples of data that the scaling exponent is in fact close to 0.75 (Armstrong, 1982, 1983, 1985; Bauchot, 1978; Eisenberg, 1981; Eisenberg and Wilson, 1978; Martin, 1981, 1982; Martin and Harvey, 1975). The coincidence between the exponent value for the scaling of brain size and that for the scaling of basal metabolic rate (0.75 in both cases) suggests the possibility of a link between the two variables. To put it another way, brain size in placental mammals scales isometrically with basal metabolic rate, such that there is, overall, a direct proportionality between the two parameters. It is therefore reasonable to seek a functional explanation that could account for an association between brain size and basal metabolic rate.

Assuming that there is, indeed, a causal connection between brain size and metabolic rate in placental mammals, the first point is that the *direction* of that connection remains to be established. It is, in principle, conceivable that the overall size of the brain in some way influences the basal metabolic rate of a mammalian species. Alternatively, it is also possible that a given basal metabolic rate is required to support a given brain size in the adult. The first possibility was, in fact, suggested by Kestner (cited in Brody, 1945), while the second possibility has recently been championed by Armstrong (1982, 1983, 1985). The choice between these two opposing interpretations must depend upon supplementary arguments, as the correlation between basal metabolic rate and brain size does not itself indicate the direction of the link. The Kestner/Brody hypothesis is difficult to defend because it is difficult to see how the brain as a whole, rather than relatively small centres in the brain, might control metabolic turnover in the body generally. On the other hand, brain tissue is relatively expensive in energetic terms and it would, at first sight, seem reasonable to propose that the brain size that any given mammal can afford to support would depend upon its metabolic turnover.

There is, however, a serious problem with *both* of the above hypotheses. In placental mammals the scatter of points around the best-fit line is markedly greater for the scaling of brain size to body size than for the scaling of basal metabolic rate. For brain size, there is a five-fold variation on either side of the best-fit line, whereas for basal metabolic rate there is only two-fold variation on either side of the best-fit line. In other words, there is a considerable fraction of variation in brain size that is not matched by variation in basal metabolic rate. To take one extreme example, the size of the human brain is approximately five times larger than expected from the best-fit line for placental mammals generally (Martin, 1981, 1983), but the basal metabolic rate for humans coincides very closely with the value expected from the Kleiber equation (Figure 8.5). This one case is inexplicable either with the Kestner–Brody hypothesis or with the Armstrong hypothesis. With the Armstrong hypothesis, in particular, it is impossible to explain how humans can support such a large brain with a basal metabolic rate that does not depart from the condition typical for mammals generally. There is a further problem with both of these hypotheses in that the scaling relationship between brain size and basal metabolic rate does not seem to be isometric for birds and reptiles (Martin, 1981). A hypothesis invoking a direct link between metabolic turnover and the size of the adult brain would therefore apply only to mammals and not to birds and reptiles.

Partly for the reasons given above, a radically different hypothesis was advanced by Martin (1981), namely that it is the metabolic turnover of the mother that constrains the growth of the fetus and notably the growth of the fetal brain. This constraint is reflected in turn in the completed size of the adult brain. Such an *indirect* connection between maternal metabolism and adult brain size allows for the intervention of other factors, such as gestation period, lactation period, the proportion of maternal resources devoted to development of her offspring, and the degree of postnatal development of the brain. Further, it can be argued that the egg-laying habits of birds and reptiles introduce an additional indirect influence in that development of the fetal brain depends upon the metabolism of the egg and not directly on the metabolism of the mother. In this case, it is quite likely that the additional complexity involved in the link between maternal metabolism and fetal brain development will fundamentally modify the scaling relationships in comparison to mammals.

It has already been demonstrated above for primates that approximately half of the variation in daily maternal investment in overall fetal growth (W_N/G) can be explained by variation in basal metabolic rate relative to maternal body size. The remaining half of the variation can probably be attributed to variations in the partitioning of maternal resources between reproduction and other functions. In addition, gestation length, lactation period and partitioning of maternal resources during lactation will all exert influences on the completed size of the adult brain. The cumulative effect of these variables can, in principle, account for the greater scatter in the scaling of mammalian brain size as opposed to the scaling of basal metabolic rate. Thus, although the hypothesis involving an indirect link between maternal metabolism and completed brain size of course requires further testing and refinement, it does account for a wider range of results from allometric analyses than do the hypotheses of Kestner/ Brody and Armstrong.

As a final note, it must be emphasised that the hypothesis invoking a relationship between maternal metabolism and fetal brain development can involve two different kinds of constraint. It is possible to envisage a direct, proximate constraint in which maternal metabolic capacity directly limits fetal growth. Such a constraint would be particularly likely at the end of pregnancy and during lactation, when daily demands on the mother are particularly high. It is also possible, however, to postulate an indirect, ultimate constraint in that selection pressures favour particular levels of maternal investment in particular species. The proportion of maternal resources devoted to fetal growth and lactation can vary between species, depending upon the reproductive strategies involved (Ross, 1988). A case in point is provided by the tree-shrews. Although tree-shrews have relatively low basal metabolic rates (Figure 8.5, Table 8.1), they show a relatively high level of daily maternal investment in fetal growth (Table 8.2). This departs from the general pattern for primates and it can be argued that tree-shrews (which have altricial neonates, in contrast to the precocial neonates of primates) simply invest a greater proportion of maternal resources in fetal growth. Similarly, it can be argued that the unexpectedly high levels of maternal daily investment in fetal growth found in mouse and dwarf lemurs (*Microcebus* and *Cheirogaleus*), in comparison to other strepsirhine primates, represent a special adaptation in terms of the allocation of maternal resources to reproduction. With both indirect and direct constraints, however, it is reasonable to expect that for mammals generally investment in fetal growth will, overall scale in a similar fashion to basal metabolic rate. Indirect constraints will allow greater scatter around the best-fit line, reflecting the special adaptations of individual species; but the overall scaling relationship will remain the same.

CONCLUSIONS

In this contribution, the attempt has been made to highlight Julian Huxley's seminal contribution to bivariate scaling analysis and to demonstrate various developments that have taken place in interspecific studies based on this approach. Thanks to these developments, allometric analysis now represents a valuable general tool for the biologist with a wide range of potential applications. It has been shown that useful results can be obtained from empirical or theoretically-based analyses that distinguish between the scaling effects of body size (the scaling principle) and the special adaptations of individual species (represented by their residual values). It has also been shown that a number of problems remain to be resolved, notably with respect to statistical aspects of line-fitting. Here, it must be stressed that these problems are not of a purely statistical nature. There is an intimate interaction between statistical procedure and biological interpretation and continued work in this area will undoubtedly yield major benefits extending far beyond the interests of allometric analysis. Because of the combined biological and statistical problems, there is much to be gained in terms of increased clarity of understanding by working directly with biological parameters in such bivariate analyses. It is important that the biologist conducting scaling analyses should never lose sight of the raw biological data, and bivariate plots provide a means of visualising the data and seeking unexpected patterns by eye.

This leads naturally on to the vexed question of multivariate analysis. Although multivariate studies have become increasingly popular in comparative biological studies, the methods that are commonly used (e.g. canonical analysis, principal components analysis) have the severe disadvantage that the biologist rapidly loses sight of the raw data and is confronted with patterns that are not generally amenable to direct biological interpretation. In addition, it is highly likely that such studies do not generally cope adequately with the problem of scaling to body size. Because of the unresolved statistical problems attached to the determination of a best-fit line, even for a bivariate comparison, and because of the ever-present problem of outliers and grade distinctions, any approach that relies entirely on statistical abstraction from a multivariate cloud of points is almost bound to be badly flawed. For this reason, it is worth seeking an alternative method that proceeds step-by-step and permits the exercise of biological judgement and interpretation at every stage.

Given a set of parameters in a comparative study, one promising multivariate approach is to scale each parameter individually to body size and to examine each bivariate plot for outliers, grades and possible reasons for fitting a line of fixed slope. The hopefully 'size-free' residual values obtained from such carefully conducted bivariate analyses can then be examined with a suitable multivariate technique, in order to seek an overall pattern. For instance, it is possible to determine a matrix of Euclidean distance separating pairs of species on the basis of a table of residual values and to apply the technique of multidimensional scaling to condense this information into a two-dimensional diagram. As the residual values are determined with individual bivariate plots, it is relatively easy to monitor progress from the original data and to determine why a particular pattern of relationships emerges with the multidimensional scaling plot. This approach has, for instance, been applied to surface areas of the compartments of the mammalian gut (Martin *et al.*, 1985; Mac-Larnon *et al.*, 1986a, 1986b) and to dimensions of the skulls and teeth of monkeys belonging to the African guenon group (Martin and MacLarnon, 1988b). In the first case, the multidimensional plot based on gut dimensions successfully grouped species according to their dietary habits. In the second case, the multidimensional plots based on dental and cranial variables showed a close correspondence to hypotheses of phylogenetic groupings based on studies of chromosomal rearrangements, vocalisations and other features. The latter study is of particular interest in that it seems to indicate that the successful 'elimination' of size effects in a comparison between species really does seem to separate the more trivial scaling influence of body size from more fundamental differences reflecting degree of phylogenetic separation. If this is, indeed, the case, it is perhaps justifiable to conclude that allometric analyses may separate the scaling effects of common growth programmes (the focus of Huxley's original interest in the topic of allometric scaling) from novel phylogenetic developments that modify those programmes. This raises exciting possibilities for the future development of phylogenetic interpretation.

Acknowledgements

Part of the research on which this paper is based was supported by a 3-year project grant from the Medical Research Council, London. This grant permitted collection and analysis of data by Dr Ann MacLarnon, in her capacity as Research Assistant for the project. The following organisations have given generous assistance in providing data and/or material for the determination of gestation periods, neonatal body weights and neonatal brain weights: Jersey Wildlife Preservation Trust (Jersey Zoo); East Midlands Zoological Society (Twycross Zoo); Zoological Society of London (London Zoo and Whipsnade Zoo). Formulation of the ideas presented in the paper was greatly assisted by discussions at various stages with the following people: Dr Leslie Aiello, Mr Fred Brett, Professor John Eisenberg, Dr Paul Harvey, Dr Michael Hills, Dr Georgina Mace, Dr Phillip Payne, Miss Caroline Ross, Dr Ben Rudder, Dr Erika Wheeler. Particular thanks are due to Dr Ann MacLarnon for her active participation in various studies mentioned in this paper and for providing comments on the manuscript.

References

Aiello, L. C. (1981) 'Locomotion in the Miocene Hominoidea', *Symp. Soc. Stud. Hum. Biol.*, vol. 21, pp. 63–97.

Armstrong, E. (1982) 'A Look at Relative Brain Size in Mammals', *Neurosci. Lett.*, vol. 34, pp. 101–4.

Armstrong, E. (1983) 'Relative Brain Size and Metabolism in Mammals', *Science*, vol. 220, pp. 1302–4.

Armstrong, E. (1985) 'Relative Brain Size in Monkeys and Prosimians', *Amer. J. phys. Anthrop.*, vol. 66, pp. 263–73.

Atchley, W. R., Gaskins, C. T. and Anderson, D. (1976) 'Statistical Properties of Ratios', *Syst. Zool.* vol. 25, pp. 137–48.

Bauchot, R. (1978) 'Encephalization in Vertebrates: A New Mode of Calculation for Allometry Coefficients and Isoponderal Indices', *Brain Behav. Evol.*, vol. 15, pp. 1–18.

Benedict, F. G. (1938) 'Vital Energetics: A Study in Comparative Basal Metabolism', *Carnegie Inst. Wash. Pub.*, vol. 503, pp. 1–215.

Brody, S. (1945) *Bioenergetics and Growth* (New York: Rheinhold Pub. Co.).

Brody, S. and Procter, T. C. (1932) 'Relations Between Basal Metabolism and Mature Body Weight in Different Species of Mammals and Birds', *Res. Bull. Missouri Agric. Res. Stat.*, vol. 166, pp. 89–101.

Calder, W. A. (1984) *Size, Function and Life History* (Cambridge, Mass.: Harvard University Press).

Clarke, R. W. (1943) 'The Respiratory Exchange of *Tarsius spectrum*', *J. Mammal.*, vol. 24, pp. 94–6.

Conroy, G. C. (1987) 'Problems of Body-weight Estimation in Fossil Primates', *Int. J. Primatol.*, vol. 8, pp. 115–37.

Daniels, H. L. (1984) 'Oxygen Consumption in *Lemur fulvus*: Deviation from the Ideal Model.', *J. Mammal.*, vol. 65, pp. 584–92.

Dobler, H.-J. (1982) 'Temperaturregulation und Sauerstoffverbrauch beim Senegal- und Zwerggalago (*Galago senegalensis, Galago (Galagoides) demidovii)*', *Bonn. Zool Beitr.*, vol. 33, pp. 33–59.

Dubois, E. (1897) 'Sur le rapport du poids de l'encéphale avec la grandeur du corps chez les mammifères', *Bull. Soc. Anthrop. Paris*, vol. 8, pp. 337–76.

Eisenberg, J. F. (1981) *The Mammalian Radiations* (London: Athlone Press).

Eisenberg, J. F. and Wilson, D. E. (1978) 'Relative Brain Size and Feeding Strategies in the Chiroptera', *Evolution*, vol. 32, pp. 740–51.

Elgar, M. A. and Harvey, P. H. (1987) 'Basal Metabolic Rates in Mammals: Allometry, Phylogeny and Ecology', *Funct. Ecol.*, vol. 1, pp. 25–36.

Freedman, L. (1962) 'Growth of Muzzle Length Relative to Calvaria Length in *Papio*', *Growth*, vol. 26, pp. 117–28.

Gingerich, P. D. (1977) 'Correlation of Tooth Size and Body Size in Living Hominoid Primates, with a Note on Relative Brain Size in *Aegyptopithecus* and *Proconsul*', *Amer. J. phys. Anthrop.*, vol. 47, pp. 395–8.

Gingerich, P. D. and Smith, B. H. (1985) 'Allometric Scaling in the Dentition of Primates and Insectivores', in Jungers, W. L. (ed.), *Size and Scaling in Primate Biology* (New York: Plenum Press) pp. 257–52.

Gingerich, P. D., Smith, B. H. and Rosenberg, K. (1982) 'Allometric Scaling in the Dentition of Primates and Prediction of Body Weight from Tooth Size in Fossils', *Amer. J. phys. Anthrop.*, vol. 58, pp. 81–100.

Gould, S. J. (1966) 'Allometry and Size in Ontogeny and Phylogeny', *Biol. Rev.*, vol. 41, pp. 587–640.

Harvey, P. H. and Mace, G. M. (1982) 'Comparison between Taxa and Adaptive Trends: Problems of Methodology', in King's College Sociobiology Group (ed.), *Current Problems in Sociobiology* (Cambridge: Cambridge University Press) pp. 343–61.

Harvey, P. H., Martin, R. D. and Clutton-Brock, T. H. (1987) 'Life Histories in Comparative Perspective', in Smuts, B. B., Cheney, D. L., Seyfarth, R. M., Wrangham, R. W. and Struhsaker, T. T. (eds), *Primate Societies* (Chicago: Chicago University Press) pp. 181–96.

Harvey, P.H. and Zammuto, R. M. (1985) 'Patterns of Mortality and Age at First Reproduction in Natural Populations of Mammals', *Nature. Lond.*, vol. 315, pp. 319–20.

Hills, M. (1978) 'On Ratios – A Response to Atchley, Gaskins and Anderson', *Syst. Zool.*, vol. 27, pp. 61–2.

Hunter, J., Martin, R. D., Dixson, A. F. and Rudder, B. C. C. (1979) 'Gestation and Interbirth Intervals in the Owl Monkey (*Aotus trivirgatus griseimembra*)', *Folia primatol.*, vol. 31, pp. 165–75.

Huxley, J. S. (1932) *Problems of Relative Growth* (New York: Dial Press).

Huxley, J. S. (1958) 'Evolutionary Processes and Taxonomy with Special Reference to Grades', *Upsala Univ. Arrsk.*, vol. 6, pp. 21–39.

Huxley, J. S. and Teissier, G. (1936) 'Terminology of Relative Growth', *Nature, Lond.*, vol. 137, pp. 780–1.

Huxley, T. H. (1864) *Evidence as to Man's Place in Nature* (London: Williams and Norgate).

Izard, M. K., Wright, P. C. and Simons, E. L. (1985) 'Gestation Length in *Tarsius bancanus*', *Amer. J. primatol.*, vol. 9, pp. 327–31.

Jerison, H. J. (1973) *Evolution of the Brain and Intelligence* (New York: Academic Press).

Jungers, W. L. (ed.) (1985) *Size and Scaling in Primate Biology* (New York: Plenum Press).

Kirkwood, J. K. (1985) 'The Influence of Size on the Biology of the Dog', *J. small Anim. Pract.*, vol. 26, pp. 97–110.

Kleiber, M. (1932) 'Body Size and Metabolism', *Hilgardia*, vol. 6, pp. 315–53.

Kleiber, M. (1947) 'Body Size and Metabolic Rate', *Physiol. Rev.*, vol 27, pp. 511–41.

Kleiber, M. (1961) *The Fire of Life: An Introduction to Animal Energetics* (New York: John Wiley).

Kurland, J. A. and Pearson, J. D. (1986) 'Ecological Significance of Hypometabolism in Nonhuman Primates: Allometry, Adaptation, and Deviant Diets', *Amer. J. phys. Anthrop.*, vol. 71, pp. 445–57.

Leutenegger, W. (1973) 'Maternal-fetal Weight Relationships in Primates', *Folia primatol.*, vol. 20, pp. 280–93.

MacLarnon, A. M., Chivers, D. J. and Martin, R. D. (1986a) 'Gastrointestinal Allometry in Primates and Other Mammals including New Species', in Else, J. G. and Lee, P. C. (eds), *Primate Ecology and Conservation* (Cambridge: Cambridge University Press) pp. 75–85.

MacLarnon, A. M., Martin, R. D., Chivers, D. J. and Hladik, C. M. (1986b) 'Some Aspects of Gastrointestinal Allometry in Primates and other Mammals', in Sakka, M. (ed.) *Définition et Origines de l'Homme* (Paris: Éditions du CNRS) pp. 293–302.

Martin, R. D. (1980) 'Adaptation and Body Size in Primates', *Z. Morph. Anthrop.*, vol. 71, pp. 115–24.

Martin, R. D. (1981) 'Relative Brain Size and Metabolic Rate in Terrestrial Vertebrates', *Nature. Lond.*, vol. 293, pp. 57–60.

Martin, R. D. (1982) 'Allometric Approaches to the Evolution of the Primate Nervous System', pp. 39–56 in Armstrong, E. and Falk, D. (eds), *Primate Brain Evolution* (New York: Plenum Press).

Martin, R. D. (1983) *Human Brain Evolution in an Ecological Context* (52nd James Arthur Lecture on the Evolution of the Human Brain) (New York: American Museum of Natural History).

Martin, R. D. (1984a) 'Scaling Effects and Adaptive Strategies in Mammalian Lactation', *Symp. zool. Soc. Lond.*, vol 51, pp. 87–117.

Martin, R. D. (1984b) 'Body Size, Brain Size and Feeding Strategies', in Chivers, D. J., Wood, B. A. and Bilsborough, A. (eds), *Food Acquisition and Processing in Primates* (New York: Plenum Press) pp. 73–103.

Martin, R. D., Chivers, D. J., MacLarnon, A. M. and Hladik, C. M. (1985) 'Gastrointestinal Allometry in Primates and Other Mammals', in Jungers, W. L. (ed.), *Size and Scaling in Primate Biology* (New York: Plenum Press) pp. 61–89.

Martin, R. D. and Harvey, P. H. (1985) 'Brain Size Allometry: Ontogeny and Phylogeny, in Jungers, W. L. (ed.), *Size and Scaling in Primate Biology* (New York: Plenum Press) pp. 147–73.

Martin, R. D. and MacLarnon, A. M. (1985) 'Gestation Period, Neonatal Size and Maternal Investment in Placental Mammals', *Nature. Lond.*, vol. 313, pp. 220–23.

Martin, R. D. and MacLarnon, A. M. (1986) 'Maternal Investment in Mammals: Reply to Zeveloff and Boyce', *Nature. Lond.*, vol. 321, p. 538.

Martin, R. D. and MacLarnon, A. M. (1988a) 'Comparative Quantitative Studies of Growth and Reproduction', *Symp. zool. Soc. Lond.*, vol. 60, pp. 39–80.

Martin, R. D. and MacLarnon, A. M. (1988b) 'Quantitative Comparisons of the Skull and Teeth in Guenons', in Gautier-Hion, A., Bourlière, F., Gautier, J.-P. and Kingdon, J. *A Primate Radiation: Evolutionary Biology of the African Guenons* (Cambridge University Press: Cambridge) pp. 160–83.

Martin, R. D. and MacLarnon, A. M. (in press) 'Reproductive Patterns in Primates and Other Mammals: The Dichotomy Between Altricial and Precocial Offspring', in DeRousseau, C. J. and Morbeck, M. E. (eds), *Primate Life History and Evolution* (Alan Liss: New York).

McCormick, S. A. (1981) 'Oxygen Consumption and Torpor in the Fat-tailed Dwarf Lemur (*Cheirogaleus medius*): Rethinking Prosimian Metabolism', *Comp. Biochem. Physiol.*, vol. 68A, pp. 605–10.

McMahon, T. A. and Bonner, J. T. (1983) *On Size and Life* (New York: Scientific American Books).

McNab, B. K. (1978) 'Energetics of Arboreal Folivores: Physiological Problems and the Ecological Consequences of Feeding on an Ubiquitous Food Supply', in Montgomery, G. G. (ed.), *The Ecology of Arboreal Folivores* (Smithsonian Institution Press: Washington) pp. 153–62.

McNab, B. K. (1980) 'Food Habits, Energetics and the Population Biology of Mammals', *Amer. Nat.*, vol. 116, pp. 106–24.

McNab, B. K. (1983) 'Energetics, Body Size and the Limits to Endothermy', *J. Zool., Lond.*, vol. 199, pp. 1–29.

McNab, B. K. (1986) 'The Influence of Food Habits on the Energetics of Eutherian Mammals', *Ecol. Monogr.*, vol. 56, pp. 1–19.

Müller, E. F. (1983) 'Wärme- und Energiehaushalt bei Halbaffen (Prosimiae)', *Bonn. zool. Beitr.*, vol. 34, pp. 29–71.

Müller, E. F. (1985) 'Basal Metabolic Rate in Primates – the Possible Role of Phylogenetic and Ecological Factors', *Comp. Biochem. Physiol.*, vol. 81A, pp. 707–11.

Müller, E. F., Kamau, J. M. Z. and Maloiy, G. M. O. (1983) 'A Comparative Study of Basal Metabolism and Thermoregulation in a Folivorous

(*Colobus guereza*) and an Omnivorous (*Cercopithecus mitis*) Primate Species', *Comp. Biochem. Physiol.*, vol. 74A, pp. 319–22.

Müller, E. F., Nieschalk, U. and Meier, B. (1985) 'Thermoregulation in the Slender Loris (*Loris tardigradus*)', *Folia primatol.*, vol. 44, pp. 216–26.

Nagy, K. A. and Milton, K. (1979) 'Energy Metabolism and Food Consumption by Wild Howler Monkeys (*Alouatta palliata*)', *Ecology*, vol. 60, pp. 475–80.

Payne, P. R. and Wheeler, E. F. (1967) 'Comparative Nutrition in Pregnancy', *Nature. Lond.*, vol. 215, pp. 1134–6.

Peters, R. H. (1983) *The Ecological Implications of Body Size* (Cambridge: Cambridge University Press).

Pézard, A. (1918) 'Le conditionnement physiologique des caractères sexuels secondaires chez les oiseaux', *Bull. Biol. France Belg.*, vol. 52, pp. 1–76.

Portmann, A. (1938) 'Die Ontogenese der Säugetiere als Evolutionsproblem. II. Zahl der Jungen. Tragzeit und Ausbildungsgrad der Jungen bei der Geburt', *Biomorphosis*, vol. 1, pp. 109–26.

Portmann, A. (1939) 'Nesthocker und Nestflüchter als Entwicklungszustände von verschiedener Wertigkeit bei Vögeln und Säugern', *Rev. suisse Zool.*, vol. 46, pp. 385–90.

Portmann, A. (1941) 'Die Tragzeit der Primaten und die Dauer der Schwangerschaft beim Menschen: Ein Problem der vergleichenden Biologie', *Rev. suisse Zool.*, vol. 48, pp. 511–18.

Portmann, A. (1965) 'Über die Evolution der Tragzeit bei Säugetieren', *Rev. suisse Zool.*, vol, 72, pp. 658–66.

Rasmussen, D. T. and Izard, M. K. (1988) 'Scaling of Growth and Life History Traits Relative to Body Size, Brain Size, and Metabolic Rate in Lorises and Galagos (Lorisidae, Primates)', *Amer. J. phys. Anthrop.*, vol. 75, pp. 357–67.

Reeve, E. C. R. and Huxley, J. S. (1945) 'Some Problems in the Study of Allometric Growth', in Le Gros Clark, W. E. and Medawar, P. B. (eds), *Essays on Growth and Form* (Oxford: Oxford University Press) pp. 121–56.

Richard, A. E. and Nicoll, M. E. (1987) 'Female Social Dominance and Basal Metabolism in a Malagasy Primate, *Propithecus verreauxi*', *Amer. J. Primatol.*, vol. 12, pp. 309–14.

Roberts, M. (1988) 'Management and Husbandry of the Western Tarsier (*Tarsius bancanus*) at the National Zoological Park', *Lab. Prim. Newsl.*, vol. 27, pp. 1–4.

Rollinson, J. and Martin, R. D. (1981) 'Comparative Aspects of Primate Locomotion, with Special Reference to Arboreal Cercopithecines', *Symp. zool. Soc. Lond.*, vol. 48, pp. 377–427.

Ross, C. (1988) 'The Intrinsic Rate of Natural Increase and Reproductive Effort in Primates', *J. Zool., Lond.*, vol. 214, pp. 199–219.

Rudder, B. C. C. (1979) *The Allometry of Primate Reproductive Parameters.* Ph.D. Thesis, University of London.

Schmidt-Nielsen, K. (1984) *Scaling: Why is Animal Size So Important?* (Cambridge: Cambridge University Press).

Shea, B. T. (1981) 'Relative Growth of the Limbs and Trunk of the African Apes', *Amer. J. phys. Anthrop.*, vol. 56, pp. 179–202.

Shea, B. T. (1985) 'Ontogenetic Allometry and Scaling: A Discussion Based on the Growth and Form of the Skull in African Apes', in Jungers, W. L. (ed.), *Size and Scaling in Primate Biology* (New York: Plenum Press) pp. 175–208.

Smith, R. J. (1985) 'The Present as a Key to the Past: Body Weight of Miocene Hominoids as a Test of Allometric Methods for Paleontological Inference', in Jungers, W. L. (ed.), *Size and Scaling in Primate Biology* (New York: Plenum Press) pp. 437–48.

Snell, O. (1891) 'Die Abhängikeit des Hirngewichtes von dem Körpergewicht und den geistigen Fähigkeiten', *Arch. Psychiatr. Nervenkr.*, vol. 23, pp. 436–46.

Sutherland, W. J., Grafen, A. and Harvey, P. H. (1986) 'Life History Correlations and Demography', *Nature. Lond.*, vol. 320, p. 88.

Teissier, G. (1931) 'Recherches morphologiques et physiologiques sur la croissance des insectes', *Trav. Stat. biol. Roscoff*, vol. 9, pp. 27–238.

Teissier, G. (1948) 'La relation d'allométrie: Sa signification statistique et biologique', *Biometrics*, vol. 4, pp. 14–48.

Teissier, G. (1955) 'Sur la détermination de l'axe d'un nuage rectiligne de points', *Biometrics*, vol. 11, pp. 344–56.

Teissier, G. (1960) 'Relative Growth', in Waterman, T. H. (ed.), *The Physiology of Crustacea* (New York: Academic Press) pp. 537–60.

Thompson, D'Arcy W. (1917) *Growth and Form* (Cambridge: Cambridge University Press).

9 The Ascent of Man

M. H. Day

Without evolutionary theory the facts of biology are sterile. Without the application of that theory to the peculiar case of man, one of the central questions of life has no rational answer. Julian Huxley was a biologist who knew that the description and elucidation of biological mechanisms, of behavioural patterns, of structure, of function and of inheritance was not the whole story of life on earth. There had to be more and it had to relate to man. He envisaged biological evolution and human evolution as two phases of a single process separated by a 'critical point' after which the properties of the evolving material underwent radical change with convergence showing dominance over divergence. This led to his concept of 'Scientific Humanism' published as an essay in which humanity was first described as a phenomenon to be studied and analysed by scientific means. It is for this reason that I will attempt to review two aspects of human evolutionary studies as a contribution to this symposium in his honour.

The two aspects of the study of human evolution that command attention at the present time concern the australopithecine ape-men, their relationships to each other and to the human line, and the second concerns *Homo erectus*; is it a true species, is it a palaeospecies linking *Homo habilis* to *Homo sapiens*, or is it an extinct side-branch of the genus *Homo*? A simple phylogenetic representation (Figure 9.1) can set out the major phases of hominid evolution in terms of sequence and chronology. It covers almost 4 million years and depicts the australopithecine, hominine and sapient phases of human evolution with two major side branches that became extinct; a simplified view but one that may set the scene for those areas of interest and controversy of today.

Australopithecine ape-men of one kind or another are now known to have existed in Africa from about 4 million years BP up to about 1 million years BP in what was believed to be two basic types – one small, gracile and generalised in form while the other was larger, more robust and more specialised in terms of dental and cranial structure. In postcranial and locomotor terms they were – and still are – hard to distinguish yet both are accepted widely as bipeds. The

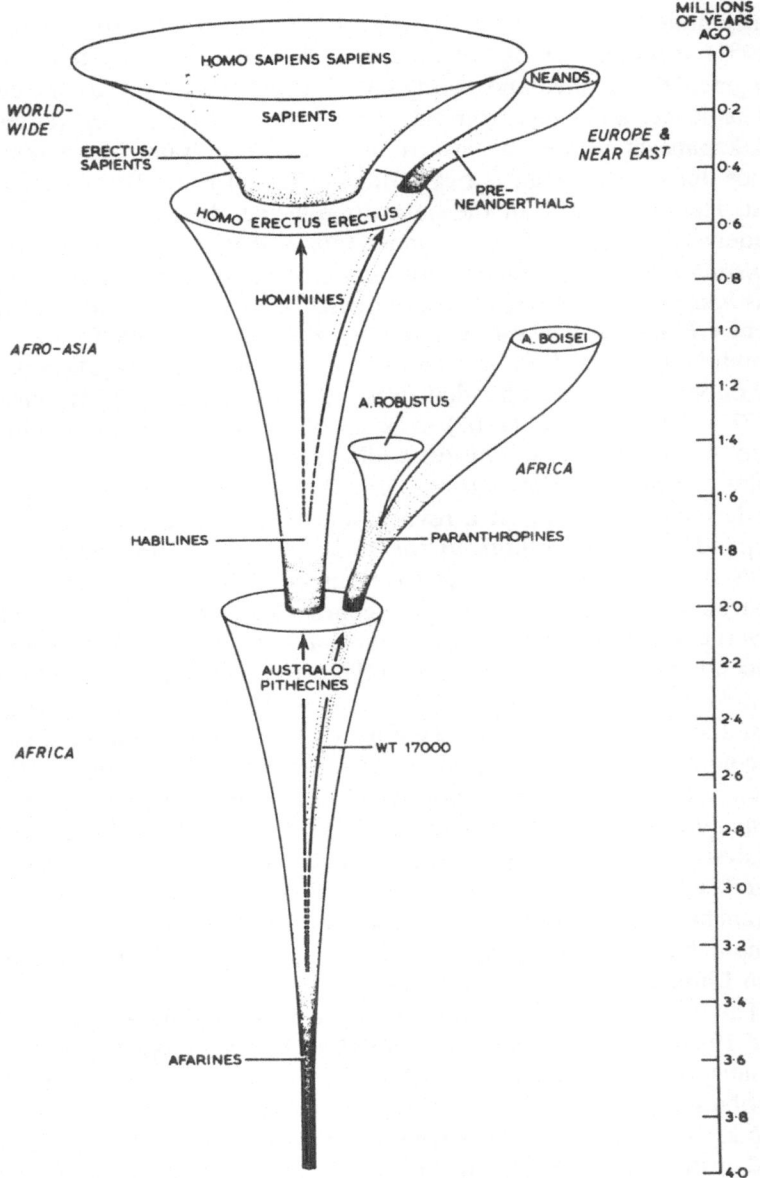

Figure 9.1   A phylogenetic representation that sets out the major phases of hominid evolution in sequence and chronology.

early finds of Dart (1925, 1948), Broom (1936, 1949) and Robinson (1953) from the Transvaal limestone cave sites were often badly fragmented, crushed and poorly dated within the Lower and Middle Pleistocene, and came from the classic sites of Taung, Sterkfontein, Makapansgat as well as Swartkrans and Kromdraai. More recent finds derive from East Africa (Olduvai, Koobi Fora, West Turkana and Hadar). Today all the australopithecines from these sites are widely regarded as from at most four species within one genus, *Australopithecus africanus*, *Australopithecus robustus*, *Australopithecus boisei* and *Australopithecus afarensis*. A second australopithecine genus *Paranthropus* was created by Broom (1938, 1949) for the hominid remains from Kromdraai and some from Swartkrans, a taxonomic arrangement that was supported later by Robinson (1972). Later Tobias (1967) sank all of the *Paranthropus* material into *Australopithecus robustus* while retaining *A. boisei* for hyper-robust forms from East Africa. *Paranthropus* as a genus has recently undergone something of a revival since cladistic analysis has been applied by some authors to the taxonomy of the early hominids (Olson, 1981; Dean, 1987). It was Rensch (1959) who introduced the concepts of 'anagenesis' and 'kladogenesis' in terms of successive and divergent evolutionary change, after many years of work during the Second World War, and Huxley (1958) the concept of 'grade' welcomed by Mayr (1963) as a 'felicitous' word. Cladistic analysis has resulted from the work of Hennig (1966); it is a system that is becoming widely, if somewhat indiscriminantly, used by modern primatologists and anthropologists. One reason for the revival of the genus *Paranthropus* would seem to be that the focus of cladistic analysis on new features, however small, variable, or of unknown significance in terms of selective advantage, leads inevitably to 'branching' of the cladogram that summarises the conclusions. It is a simple, but often totally unjustified, step to translate cladistic analysis into Linnean taxonomic decisions.

The work of Mary and Louis Leakey at Olduvai Gorge in the 1950s and 1960s produced evidence of an East African hyper-robust form of australopithecine (*A. boisei*) and a smaller creature that was attributed controversially at that time to the genus *Homo*. *Homo habilis* is represented by the type mandible (OH 7), a skull (OH 24) and postcranial bones that include a juvenile hand and an adult foot. The evidence of these bones permitted Leakey *et al.* (1964) to claim congeneric status with man for the Olduvai find. The recognition of the bipedal nature of the gait of this early member of our genus (Day and Napier, 1964) was unexpected since the Olduvai deposits that contained *Homo habilis* are reliably dated at 1.75 million years BP.

The work of Richard Leakey, in the 1970s at Koobi Fora (East of Lake Turkana) enlarged the sample of hyper-robust australopithecine remains in East Africa and added to the *Homo habilis* collection known from Olduvai by the recovery of ER 1470 from deposits of similar date. Later the recovery of Sts 53 from Sterkfontein by Hughes and Tobias (1977) not only increased the sample of this important taxon but also widened the range to include South Africa. Doubts about the validity of *Homo habilis* as a taxon have remained (Stringer, 1986) but the find by Johanson *et al.* (1987) of a partial skeleton at Olduvai Gorge, and its attribution to *Homo habilis* may well do much to clarify the situation. They find cranial and dental similarities to *Homo habilis* in their new specimen (OH 62) but postcranial similarities to the early australopithecine material from Ethiopia known as *A. afarensis*. From this it seems that *Homo habilis* is strengthened as a taxon by the addition of new material of this quality but its relationship to *Australopithecus* is underlined to the extent that the two genera solved the mechanical problems of bipedalism in very similar, if not completely identical, ways.

The recovery of hominid fossils from Hadar in Ethiopia by Johanson and his colleagues in the 1970s, and their designation as *A. afarensis* was also controversial, Johanson maintaining that all the material from this site, as well as that recovered from Laetoli, Tanzania, by Mary Leakey, belonged to one species despite some major differences in size and cranial morphology (Johanson and White, 1979). The new taxon is regarded as a stem form from over 3 million years before the present (BP) and giving rise to all subsequent hominids. Indeed, fossils from nearby by Maka and Belohdelie in Middle Awash, Ethiopia, found by White (Clark *et al.*, 1984; White, 1984) from deposits dated at 4 million years BP, has typical australopithecine femoral characteristics that are echoed by Olduvai Hominid 62 over 2 million years later. A reconstruction of the *A. afarensis* skull and some details of the skull base have led some workers to suggest that two taxa were present at Hadar rather than attributing the differences present to sexual dimorphism. The implication of this is that there was a very early split in the hominid line with one branch leading to a robust lineage and the other to a more gracile line of creatures that may have led to man (Olson, 1985; Zihlman, 1985). Others, like Tobias (1980), do not accept *A. afarensis* as a valid taxon and place all of the early material into *A. africanus*.

In summary four schemes of hominid phylogeny (Figure 9.2) were widely supported up until 1985; all four agreed that the hominid line was single in East Africa between 4 and 5 million years BP, that this line divided into two or three, one of which became extinct at about 1 million years BP after increasing specialisation, and one line gave rise to the genus *Homo*. Suddenly, in 1986, these ideas became outdated. The discovery by Richard Leakey and his colleague, Alan Walker, of a 2.5 million-year-old hyper-robust australopithecine skull (WT 17000) from west of Lake Turkana in Northern Kenya, and attributed to *A. boisei*, dramatically altered the position. It is a hyper-robust skull with massive cranial cresting, at 410 ml it has the smallest cranial capacity of any known hominid, it has a large face, evidence of big teeth and it is three quarters of a million years older than any other robust australopithecine (Walker *et al.*, 1986).

It has been widely assumed for many years that the robust australopithecines became more 'robust' and specialised as the Plio-Pleistocene progressed, eventually becoming so specialised that they became extinct. This view is no longer tenable since the earliest known example has the smallest cranial capacity and the largest cranial cresting known for hominids. It has also been suggested that *A. robustus* (a smaller form from South Africa) was ancestral to *A. boisei* (Rak, 1985). This is now unlikely since WT 17000 is clearly of the *A. boisei* lineage and is related to the most recent example of that species. *A. robustus* is thus either a related smaller species derived from a much earlier ancestral form or has evolved independently from *A. africanus* in Southern Africa.

Following the discovery of WT 17000 two of the four phylogenetic schemes (Figure 9.2) (Johanson/White and Olson) have been withdrawn and have been replaced by revised ideas (Figure 9.3a). The first of the new schemes (Walker and Johanson in Delson, 1987) sees WT 17000 as a clear hyper-robust australopithecine (*A. boisei*) derived from a single stem form *A. afarensis*. The remaining relationships are far less clear. They are uncertain as to whether both *A. africanus* and the genus *Homo* are derived from *A. afarensis* yet feel that *A. robustus* is derived from *A. africanus*. Finally this model is also uncertain whether or not the genus *Homo* is derived from *A. africanus*. This approach is so cautious that it only asserts that *A. boisei* has a long and continuous evolutionary history from 2.5 my BP to 1 my BP. The second scheme (Figure 9.3b) (Delson, Grine, Howell and Olson in Delson, 1987) divides *A. afarensis* into two phylogenetic lines, one leading through *A. africanus* to the genus

147

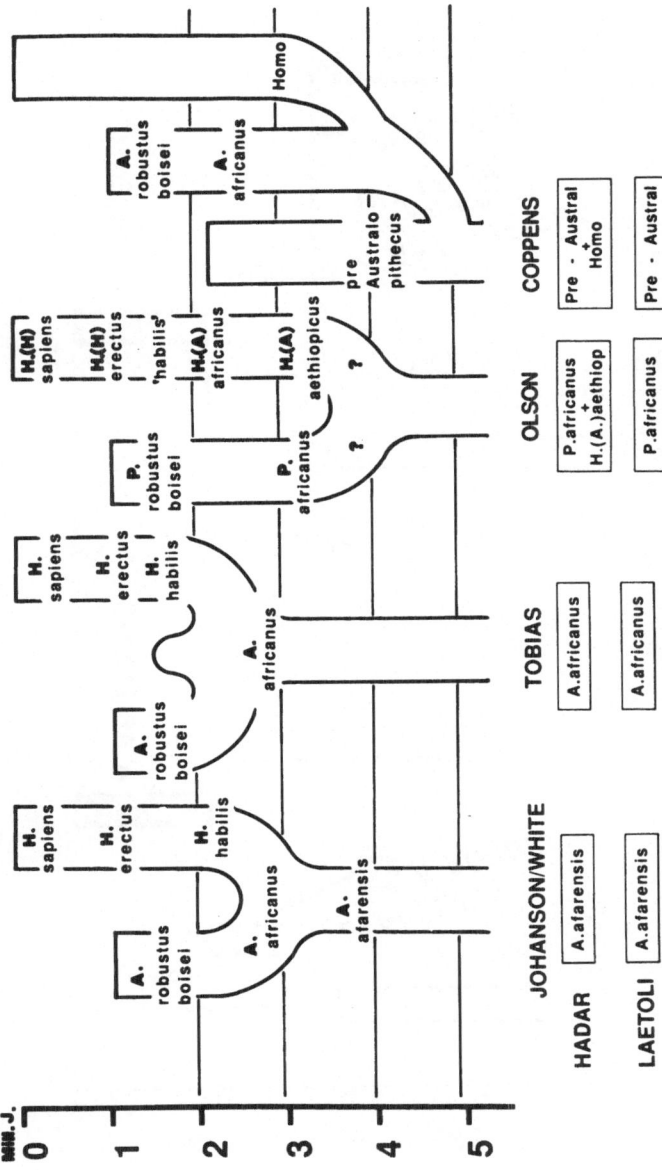

Figure 9.2 Four schemes of hominid phylogeny widely supported until 1985 (After Bräuer).

148

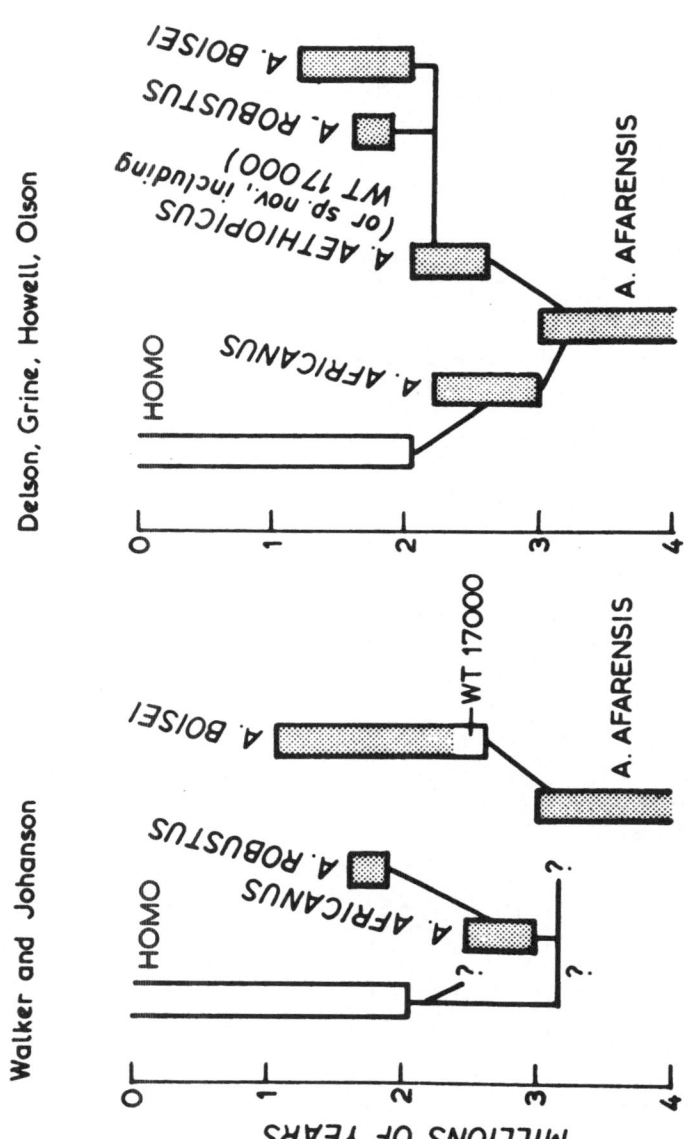

Figure 9.3   The hominid phylogenetic scheme favoured by Walker and Johanson and that favoured by Delson, Grine, Howell and Olson (redrawn after Delson, 1987).

Homo the other through WT 17000 to *A. boisei* and *A. robustus* as two distinct terminal forms. WT 17000 is seen as a new species (*'Australopithecus aethiopicus'*) at the root of the robust clade. (This scheme has much in common with that of Skelton *et al.*, 1986). Both schemes have serious drawbacks. The first suggests that 'robust' features appear in two phyletic lines, not one as might be expected. Secondly, the recognition of *A. africanus* as a sister species or ancestor of *Homo (? habilis)* requires an explanation of why there are no accredited examples of *A. africanus* from East Africa, or members of the genus *Homo* at earlier than 2 million years from anywhere? The second scheme is also contested since it requires a new species on the basis of one cranial specimen. It is favoured by those who support the genus *Paranthropus* since generic separation occurs at 3 million years and indeed *A. africanus* could be included in the genus *Homo*; an old idea given new life by cladistic analysis.

It is a matter of some irony that after over 60 years the australo-pithecines are still a puzzle in terms of their relationships to each other and to the human line. However, some facts are now firmly established, such as the clear and unequivocal evidence of bipedalism in early hominids well ahead of brain expansion and that manipula-tive ability leading to toolmaking expanded greatly at about the time of the first evidence of the genus *Homo* in East Africa.

If we turn to *Homo erectus* we can also find that what was a seemingly clear cut taxon is now under intense scrutiny both in its possible origins and in its fate. What is now known as *Homo erectus* originated as the discoveries of Eugene Dubois in Java at the turn of the century. The Java man skull-cap and femur led to the designation *Pithecanthropus erectus* (Dubois, 1894) from the belief that the skull showed features that were intermediate between those of ape and man, and the anatomical features of the femur that were similar to those of modern man. It is once again ironic that the modern human, bipedal features of this femur that were confirmed by many anatom-ists (Manouvrier, 1895; Hepburn, 1897; Weinert, 1928; Weidenreich, 1941; Le Gros Clark, 1964; Day and Molleson, 1973; Lamy, 1984) are probably due to the fact that the femur is indeed modern human and intrusive in the Middle Pleistocene deposits of Trinil, Java, where it was found (Day, 1984). Later finds from China (Black, 1931), North Africa (Arambourg, 1954) and Sangiran, Java (von Koenigswald, 1938) confirmed the cranial features of *P. erectus* but not the postcranial features as shown by the Trinil femur.

During the past 20 years the picture has changed radically. Most of the early finds are grouped in one taxon as *Homo erectus* by most modern authorities, and many new finds have been made, some of which are quite spectacular. The new material derives from East Africa (Olduvai Gorge, Koobi Fora, West Turkana), North Africa (Salé), Northern Europe (Bilzingsleben) and India (Narmada). Some other sites such as Arago, Mauer, Petralona and Vértesszöllös are regarded by some as having produced material attributable to *Homo erectus* although there is no clear agreement about all these sites.

The features of the material that has in the past given *Homo erectus* taxonomic unity include a long low cranial vault, flattened parietal and frontal bones, an angulated occiput, a strongly marked and continuous supraorbital torus, a small mastoid process, a supra-mastoid crest and a torus angularis on the parietal bone. The vault also shows a parietal sagittal keel with parasagittal flattenings and a tent-shaped coronal section with a low maximum breadth between the supramastoid crests. The vault bones are thick and the inion and the opisthocranion coincide. The mandible is chinless with a thick body, a large bicondylar breadth and often multiple mental foramina. Dentally the shovel-shaped incisors have basal tubercles, the molars and the premolars have a cingulum, the molars have a dryopithecine cusp pattern, tending to the 'plus' pattern by metaconid reduction. The femora are platymeric and thick walled and have a low narrow point. The pelvic structure is also distinctive in that the acetabulum is large, there are stout acetabulocristal buttresses and medially rotated ischial tuberosities. The recognition of the femoral features (Weiden-reich, 1941) and the pelvic features (Day, 1971) as being distinctive of *Homo erectus* has disposed of an earlier view that *Homo erectus* was the same as *Homo sapiens* in its postcranial features.

This group of about thirty features would seem to be sufficient to characterise the taxon *Homo erectus* in its fully developed form. The original finds from Java have been added to in recent years but without much clarification of the picture. In terms of dating (Pope, 1985) and fauna (Groves, 1985) there seems to be confusion; it may be, however, that none of the hominid material is more than 1.3 million years old (Pope and Cronin, 1984). Many taxa have been proposed for the Javan material but most authorities would place all of it in *Homo erectus*. The other material from the Far East derives from China. In 1980 the recovery of a calvarium of *Homo erectus* from Hexian confirms the presence of this form in China at between 240–280 000 years BP (Wu and Dong, 1982).

The recovery of a *Homo erectus* skull from Salé in North African deposits dated at about 400 000 years BP was an important event (Jaeger, 1975, 1981) that paralleled the recovery of the Arago remains from southern France from deposits now believed to be of about the same age (de Lumley and de Lumley, 1971). Also from Europe there is the Bilzingsleben skull which has many *Homo erectus* features and is dated at about 228 000 years BP (Vlček, 1978, 1983a, b).

East Africa has produced not only the well known Olduvai Hominid 9 *Homo erectus* calvaria at about 1 million years BP but also ER 3733 and ER 3883, two older skulls at about 1.8 million years BP, skulls whose contemporaneity with the robust australopithecine ER 406 sank the 'Single Species Hypothesis' of hominid evolution and opened the way for further phylogenetic speculation. By far the most exciting and complete specimen of a hominid ever recovered comes from a site to the West of Lake Turkana (formerly Lake Rudolf) and was found by Richard Leakey and his team in 1984 and 1985 (Brown *et al.*, 1985). It is the skeleton of a boy that is almost complete and with a full dentition. Estimates of his age suggest about 12 years and a stature of about 5'3'' – 5'7''. The site is reliably dated at 1.6 million years BP. The initial description leaves little doubt that the specimen should be attributed to *Homo erectus* on cranial and dental grounds; postcranially it is persuasive also since he has a suite of features already described earlier for the femur and pelvis of *Homo erectus* (Day, 1971, 1982). The three sites involved include Olduvai Gorge, Bed IV (OH 28); Koobi Fora (ER 3228) and Arago XLIV) whose time range covers one million years and from Tanzania to France. They have identical but unusual features. If they are not specific hallmarks of *Homo erectus* then at least we can say, as we did with the australopithecines, that these hominids solved the problems of bipedality and upright posture in the same way and that the solution lasted for a least one million years.

The first example of *Homo erectus* to be recovered from the Indian subcontinent was found by Arun Sonakia in 1982 at Hathnora in the Narmada valley, Madhya Pradesh (de Lumley and Sonakia, 1985). It has many typically *Homo erectus* features but also some that are advanced such as a tall cranial vault and a cranial capacity that may lie between 1155–1420 cc. It is associated with an Upper Acheulean industry with both hand-axes and cleavers.

This brief survey of the old and the new material attributed to *Homo erectus* draws attention to some of the problems of this taxon highlighted earlier (Day, 1986). Does *Homo erectus* exist as a true species or should it be sunk into *Homo sapiens*? Is it an extinct

species that had no part in the evolution of *Homo sapiens*? Does
Homo erectus exist in the fossil record of Europe? Are the Asian
specimens of *Homo erectus* the only true examples of this species?
Should *Homo erectus* be regarded as a palaeospecies that exists in
classical form as a segment of the line that emerged from *Homo
habilis* and gave rise to *Homo sapiens*? These questions have been
answered in various ways by differing authors. Jelinek (1978, 1981)
and Thoma (1973) have argued that *Homo erectus* should be sunk
into *Homo sapiens* while many others including Rightmire (1984)
maintain that *Homo erectus* is a good species. Howell (1981, 1986)
and Stringer (1981) have questioned the existence of a true example
in Europe whereas Jaeger (1975) and Vlček (1978, 1983a, b) are in no
doubt that their material belongs to *Homo erectus*.

The morphological correspondence of groups of homologous structures forms the basis of the well established comparative method of
taxonomy that results in classification by similarity and the recognition of affinity. It is a method that encourages the identification of
natural groupings down to species level. In the case of *Homo erectus*
the list of cranial features that are widely recognised has been
supplemented by the postcranial features so that together the morphological comparisons are now wider than ever before.

Definitions of *Homo erectus* based on combinations of features
have been given by many authors (Weidenrich, 1943; Le Gros Clark,
1964; Howell, 1978; Howells, 1980; Day and Stringer, 1982; Rightmire, 1984) with varying degrees of success.

The cladistic system of analysis has been taken up somewhat
belatedly by anthropologists but is now being applied with proselyte
fervour by a number of practitioners. The cladistic system demands
that if a group of new characters (or even a single new character) can
be found to be shared between two forms then they must be related
to the exclusion of others that do not share this feature or features.
(Indeed to the purist only one will convict). Attempts have been
made to identify new or derived characters that are unique to *Homo
erectus* (Andrews, 1984; Wood, 1984; Bilsborough and Wood, 1986;
Hublin, 1986). This has led Andrews (1984) to the conclusion that
Homo erectus was confined to Asia since he has been unable to
identify derived features in material from outside that continent. The
weakness of this approach is apparent when it is noted that of his
seven features said to be 'diagnostic of the species' (*Homo erectus*) at
least four can be found in African hominid fossils.

A more widely held view would see *Homo erectus* as a palaeospecies that shows evidence of evolution through time from a more primitive ancestor such as *Homo habilis* to a more advanced successor such as *Homo sapiens* (Le Gros Clark, 1964; Campbell, 1972; Wolpoff, 1980; Howells, 1981; Day, 1984, 1986). Whether fossils such as ER 3773 and 3883 and WT 15000 are finally judged as *habilis/erectus* transitionals, and Omo II, Arago and Petralona as *erectus/sapiens* transitionals depends on one's view of evolution and of the mosaic evolutionary process. If the attributions of the finders are accepted then *Homo erectus* spans about one and a quarter million years from 1.6 million BP to perhaps 250 000 BP; secondly his range extends from the Far East through India to Europe, the oldest finds being from Africa and suggesting an African origin. Those who support the punctuational model of human evolution see *Homo erectus* as a true species that varied little during its long existence.

Homo erectus is therefore proving to be as much a puzzle as the australopithecines: in both cases neither do we know precisely from what they arose nor to what they gave rise. But we do know that they lived, breathed, reproduced and evolved and that they were of our own kind. For this reason, if no other, they demand the attention of all students of man as they did of Julian Huxley.

References

Andrews, P. (1984) 'An Alternative Interpretation of the Characters used to Define *Homo erectus*', *Courier Forschunginstitut Senckenberg*, vol. 69, pp. 167–75.

Arambourg, C. (1954) 'L'hominien fossile de Ternifine (Algérie)', *Comptes Rendus Academie Science Paris*, vol. 239, pp. 893–95.

Bilsborough, A. and Wood, B. A. (1986) 'The Nature, Origin and Fate of *Homo erectus*', in B. Wood, L. Martin and P. Andrews (eds), *Major Topics in Primate and Human Evolution* (Cambridge: Cambridge University Press).

Black, D. (1931) 'On an Adolescent Skull of *Sinanthropus pekinensis* in Comparison with an Adult Skull of the same Species and with other Hominid Skulls, Recent and Fossil', *Palaeontology Sinica*, Ser. D. 7, II, pp. 1–145.

Broom, R. (1936) 'A New Fossil Anthropoid Skull from South Africa', *Nature*, vol. 138, pp. 486–8.

Broom, R. (1938) 'The Pleistocene Anthropoid Apes of South Africa', *Nature*, vol. 142, pp. 377–9.

Broom, R. (1949) 'Another Type of Fossil Ape-man', *Nature*, vol. 163, p. 57.

Brown, F. H., Harris, J., Leakey, R. E. and Walker, A. (1985) 'Early *Homo erectus* Skeleton from West Lake Turkana, Kenya', *Nature*, vol. 316, pp. 788–92.

Campbell, B. G. (1972) 'Conceptual Progress in Physical Anthropology: Fossil Man', *Annual Review of Anthropology*, vol. 1, pp. 27–54.

Clark, J. D. *et al.* (1984) 'Palaeoanthropological discoveries in the Middle Awash Valley, Ethiopia', *Nature*, vol. 307, pp. 423–8.

Clark, W. E. Le Gros (1964) *The Fossil Evidence for Human Evolution* (Chicago: University of Chicago Press).

Dart, R. A. (1925) '*Australopithecus africanus*: The man-ape of South Africa', *Nature*, vol. 115, pp. 195–9.

Day, M. H. (1971) 'Postcranial Remains of *Homo erectus* from Bed IV, Olduvai Gorge, Tanzania', *Nature*, vol. 232, pp. 383–7.

Day, M. H. (1982) 'The *Homo erectus* pelvis: Punctuation or Gradualism?', *1st International Congress for Human Palaeontology, Nice* (Prétirage) pp. 411–21.

Day, M. H. (1984) 'The Postcranial Remains of *Homo erectus* from Africa, Asia and Possibly Europe', *Courier Forschunginstitut Senckenberg*, vol. 69, pp. 113–21.

Day, M. H. (1986) 'Bipedalism: Pressures, Origins and Modes', in B. Wood, L. Martin and P. Andrews (eds), *Major Topics in Primate and Human Evolution* (Cambridge: Cambridge University Press).

Day, M. H. and Molleson, T. I. (1973) 'The Trinil Femora', in *Human Evolution; Symposium of the Society for the Study of Human Biology*, (London: Taylor and Francis) vol. 11, pp. 127–54.

Day, M. H. and Napier, J. R. (1964) 'Hominid Fossils from Bed I, Olduvai Gorge, Tanganyika. Fossil Foot Bones', *Nature*, vol. 201, pp. 967–70.

Day, M. H. and Stringer, C. B. (1982) 'A Reconsideration of the Omo Kibish Remains and the *erectus-sapiens* Transition', *1st International Congress for Human Paleontology, Nice* (Prétirage) pp. 814–46.

Dean, M. C. (1987) 'Growth Layers and Incremental Markings in Hard Tissues; a Review of the Literature and some Preliminary Observations about Enamel Structure in *Paranthropus boisei*', *Journal of Human Evolution*, vol. 16, pp. 157–72.

Delson, E. (1987) 'Evolution and palaeobiology of Robust *Australopithecus*', *Nature*, vol. 327, pp. 654–5.

Dubois, E. (1894) '*Pithecanthropus erectus*, eine menschenaehnliche Ubergangsform aus Java' (Batavia: Landesdruckerei).

Groves, C. P. (1985) 'Plio-Pleistocene Mammals in Island Southeast Asia', *Modern Quaternary Research SE Asia*, vol. 9, pp. 43–54.

Hennig, W. (1966) *Phylogenetic Systematics* (Urbana: University of Illinois Press).

Hepburn, D. (1897) 'The Trinil Femur (*Pithecanthropus erectus*) Contrasted with the Femora of Various Savage and Civilised Races', *Journal Anatomy and Physiology*, vol. 31, pp. 1–17.

Howell, F. C. (1978) 'Hominidae', in V. J. Maglio and H. B. S. Cooke (eds), *Evolution of African Mammals* (Cambridge, Massachusetts: Harvard University Press).

Howell, F. C. (1981) 'Some Views of *Homo erectus* with Special Reference to its Occurrence in Europe', in B. A. Sigmon and J. S. Cybulski (eds), *Homo erectus: Papers in Honor of Davidson Black* (Toronto: University of Toronto Press).

Howell, F. C. (1986) 'Variabilité chez *Homo erectus*, et problème de la présence de cette espèce en Europe', *L'Anthropologie*, Paris, vol. 90(3) pp. 447–81.

Howells, W. W. (1980) '*Homo erectus* – Who, When and Where: a Survey', *Yearbook of Physical Anthropology*, vol. 23, pp. 1–23.

Howells, W. W. (1981) '*Homo erectus* in Human Descent: Ideas and Problems', in B. A Sigmon and J. S. Cybulski (eds), *Homo erectus: Papers in Honor of Davidson Black* (Toronto: University of Toronto Press).

Hublin, J.-J. (1986) 'Some comments on the diagnostic features of *Homo erectus*', Special Volume in honour of Jan Jelinek. *Anthropos (Brno)*, vol. 23, pp. 175–87.

Hughes, A. R. and Tobias, P. V. (1977) 'A Fossil Skull Probably of the Genus *Homo* from Sterkfontein, Transvaal', *Nature*, vol. 265, pp. 310–12.

Huxley, J. S. (1958) 'Evolutionary Processes and Taxonomy with Special Reference to Grades', *Uppsala universitets arsskrift*, pp. 21–39.

Jaeger, J.-J. (1975) 'Découverte d'un crâne d'Hominidé dans le Pléistocène moyen du Maroc', in *Problèmes actuels de Paléontologie – Evolution des Vertébrés* (Paris: Colloques internationaux CNRS, 218), pp. 897–902.

Jaeger, J.-J. (1975) 'Découverte d'un crâne d'Hominidé dans le Pléistocène moyen du Maroc', in *Problèmes actuels de Paléontologie – Evolution des Vertébrés* (Paris: Colloques internationaux CNRS, 218) pp. 897–902.

Jaeger, J.-J. (1981) 'Les hommes fossiles du Pleistocene moyen du Mahgreb dans leur cadre géologique, chronologique et paléoécologique', in *Homo erectus: Papers in Honor of Davidson Black* (Toronto: University of Toronto Press).

Jelinek, J. (1978) '*Homo erectus* or *Homo sapiens*?', in D. J. Chivers and K. A. Joysey (eds), *Recent Advances in Primatology* (London: Academic Press).

Jelinek, J. (ed.) (1981) '*Homo erectus* and His Time', *Anthropologie (Brno)*, vol. 19, pp. 1–96.

Johanson, D. C. *et al.* (1987) 'New Partial Skeleton of *Homo habilis* from Olduvai Gorge, Tanzania', *Nature*, vol. 327, pp. 205–9.

Johanson, D. C. and White, T. D. (1979) 'A Systematic Assessment of Early African Hominids', *Science*, vol. 202, pp. 321–30.

Koenigswald, G. H. R. von (1938) 'Ein neuer *Pithecanthropus*-Schädel', *Proceedings of the Academy of Science, Amsterdam*, vol. 41, pp. 185–92.

Lamy, P. (1984) 'Les fémurs des *Homo erectus* d'après les découvertes faites à Java', in *L'Homme Fossile et son Environnement à Java* (Paris: Muséum national d'Histoire naturelle).

Leakey, L. S. B., Tobias, P. V. and Napier, J. R. (1964) 'A new species of the genus *Homo* from Olduvai Gorge', *Nature*, vol. 202, pp. 7–9.

Lumley, H. de and Lumley, M.-A. de (1971) 'Découverte de restes humains anténéandertaliens datés du début de Riss à la Caune de l'Arago (Tautavel, Pyrénées-Orientales)', *C.r. Acad. Sci.*, Paris, vol. 272, pp. 1729–42.

Lumley, M.-A. de and Sonakia, A. (1985) 'Première découverte d'un *Homo erectus* sur le continent indien à Hathnora, dans la moyenne vallée de la Narmada', *L'Anthropologie*, Paris, vol. 89, pp. 13–61.

Manouvrier, L. (1895) 'Discussion du "*Pithecanthropus erectus*" comme précurseur présumé de l'homme', *Bulletin of the Anthropological Society, Paris*, vol. 6, pp. 12–47.

Mayr, E. (1963) *Animal Species and Evolution* (Cambridge, Massachusetts: Harvard University Press).

Olson, T. R. (1981) 'Basicranial Morphology of the Extant Hominoids and Pliocene Hominids: the New Material from the Hadar Formation, Ethiopia, and its Significance in Early Human Evolution and Taxonomy', in C. B. Stringer (ed.), *Aspects of Human Evolution* (London: Taylor and Francis).

Olson, T. R. (1985) 'Cranial Morphology and Systematics of the Hadar Formation Hominids and *"Australopithecus" africanus*', in E. Delson (ed.), *Ancestors* (New York: Alan R. Liss, Inc.).

Pope, G. G. (1985) 'Taxonomy, Dating and Paleoenvironment: the Paleoecology of the Early Far Eastern Hominids', *Modern Quaternary Research SE Asia*, vol. 9, pp. 65–80.

Pope, G. G. and Cronin, J. E. (1984) 'The Asian Hominidae', *Journal of Human Evolution*, vol. 13, pp. 377–96.

Rak, Y. (1985) 'Sexual Dimorphism, Ontogeny and the Beginning of Differentiation of the Robust Australopithecine Clade', in P. V. Tobias (ed.), *Hominid Evolution: Past, Present and Future* (New York: Alan R. Liss, Inc.).

Rensch, B. (1959) *Evolution Above the Species Level* (London: Methuen).

Rightmire, G. P. (1984) 'Comparisons of *Homo erectus* from Africa and Southeast Asia', *Courier Forschunginstitut Senckenberg*, vol. 69, pp. 83–98.

Robinson, J. T. (1953) '*Telanthropus* and its Phylogenetic Significance', *American Journal of Physical Anthropology*, vol. 11, pp. 445–501.

Robinson, J. T. (1972) *Early Hominid Posture and Locomotion* (Chicago: Chicago University Press).

Skelton, R. R., McHenry, H. M. and Drawhorn, G. M. (1986) 'Phylogenetic Analysis of Early Hominids', *Current Anthropology*, vol. 27, pp. 21–43.

Stringer, C. B. (1981) 'The Dating of European Middle Pleistocene Hominids and the Existence of *Homo erectus* in Europe', *Anthropologie (Brno)*, vol. 19, pp. 3–14.

Stringer, C. B. (1986) 'The Credibility of *Homo habilis*', in B. Wood, L. Martin and P. Andrews (eds), *Major Topics in Primate and Human Evolution* (Cambridge: Cambridge University Press).

Thoma, A. (1973) 'New Evidence for the Polycentric Evolution of *Homo sapiens*', *Journal of Human Evolution*, vol. 2, pp. 529–36.

Tobias, P. V. (1967) 'The Taxonomy and Phylogeny of the Australopithecines, in B. Chiarelli (ed.), *Taxonomy of Old World Primates with References to the Origin of Man* (Torino: Roserberg and Sellier).

Tobias, P. V. (1980) '*"Australopithecus afarensis"* and *A. africanus*: Critique and an alternative hypothesis', *African Palaeontology*, vol. 23, pp. 1–17.

Vlček, E. (1978) 'A New Discovery of *Homo erectus* in Central Europe', *Journal of Human Evolution*, vol. 7, pp. 239–51.

Vlček, E. (1983a) 'Ueber einen weiteren Schädelreste des *Homo erectus* von Bilzingsleben. 4. Mitteilung', *Ethnographie-Archaeologie Zeitschrift*, vol. 24, pp. 321–5.

Vlček, E. (1983b) 'Die Neufunde vom *Homo erectus* aus dem mittelpleistozänen Travertinkomplex', in D. H. Mai, D. Mania, T. Notzold, V.

Toepfer, E. Vlcek and W.-D. Heinrich (eds), *Bilzingleben II*, (Berlin: Veb Deutscher Verlag de Wissenschaften).

Walker, A., Leakey, R. E., Harris, J. M. and Brown, F. H. (1986) '2.5 Myr *Australopithecus boisei* from West of Lake Turkana, Kenya', *Nature*, vol. 322, pp. 517–22.

Weidenreich, F. (1941) The Extremity Bones of *Sinanthropus pekinensis*', *Palaeontologica Sinica*, New Series D, vol. 5, pp. 1–150.

Weidenreich, F. (1943) 'The "Neanderthal Man" and the Ancestors of "*Homo sapiens*"', *American Anthropology*, vol. 42, pp. 375–83.

Weinert, H. (1928) '*Pithecanthropus erectus*', *Zeitschrift Gesichte Anatomie*, vol. 87, pp. 429–547.

White, T. D. (1984) 'Pliocene Hominids from the Middle Awash, Ethiopia', *Courier Forschunginstitut Senckenberg*, vol. 69, pp. 57–68.

Wolpoff, M. H. (1980) *Paleoanthropology* (New York: Knopf).

Wood, B. A. (1984) 'The Origin of *Homo erectus*', *Courier Forschunginstitut Senckenberg*, vol. 69, pp. 99–111.

Wu, R. and Dong, X. (1982) 'Preliminary Study of *Homo erectus* Remains from Hexian, Anhul', *Acta Anthropologica Sinica*, vol. I, pp. 2–13.

Zihlman, A. L. (1985) '*Australopithecus afarensis*: Two Sexes or Two Species?', in P. V. Tobias (ed.), *Hominid Evolution: Past, Present and Future* (New York: Alan R. Liss, Inc.).

10 Human Geographical Variation

G. Ainsworth Harrison

In 1935 Julian Huxley in collaboration with A. C. Haddon published a small book entitled *We Europeans*. As a frontispiece there were head photographs of sixteen males of European origin and the reader was invited to guess their country of birth. I well remember getting none right and I suspect few readers did much better. On relooking at these photographs I suspect that they were not selected at random! But the point was well made that racial/national stereotypes are a myth, and, more positively, that human beings in every country are extremely variable.

This conclusion, based upon the characters of external appearance, whose genetic basis remains unresolved, has been more than confirmed and extended by research on variation which is more simply inherited and adequately understood: characters such as blood groups, haemoglobin, serum proteins, isoenzymes and now DNA itself. To exemplify the magnitude of within population genetic heterogeneity H. Harris (1980) calculated that on the basis of only 14 polymorphic gene loci in the British, the most *commonly* occurring combinations were to be found in only 0.06 per cent of the population and the likelihood of two randomly selected people being identical is 1:32 000! A very conservative estimate of the number of polymorphic structural genes in the human genome is 8000 so the possible combinations well exceed the number of all the people who have ever lived. This genetic uniqueness of every individual is currently being reinforced by analysis of DNA, by restriction endonuclease enzymes which detect changes in the base-pair sequences. These enzymes are revealing innumerable restriction fragment polymorphisms which are even more frequent in the non-coding regions of the DNA than in the coding ones. We now have 'genetic fingerprints' of DNA variation forcefully demonstrating the extent of our individuality.

The photographs of Huxley and Haddon also implied that most of the variation in the human species lay within single populations and that different populations, particularly neighbouring ones, were quite similar to one another. This view is also confirmed by examination of simply inherited traits. Lewontin (1982), for example, has estimated from a number of blood polymorphic systems that about 85 per cent of the total human gene frequency variance occurs in single populations of a limited locality, about 10 per cent within national or similar groups of restricted geographical range and only 5 per cent between major continental divisions of mankind. Populations are much more similar than people!

In considering, then, patterns of genetic geographic variation, we are dealing with but a small part of human variability. This, however, is not to regard it, as some people tend to do for political purposes, as trivial in nature and inconsequential in effect. Distinct patterns occur and they need to be studied, both so that their biological causes and significance can be analysed and so that we can properly understand the phenomenon of race.

A number of features characterise this variability. First, populations, and particularly neighbouring ones, tend to have the same genes present: what varies from place to place is their frequency, so that whilst a particular gene may be common in one population, it may be rare in another. Put more technically, varying genetic systems tend to be polymorphic and what varies is the level at which the polymorphism is set. It is this phenomenon which largely accounts for the fact that when the variance is partitioned into hierarchies most of it is attributed to single populations.

Monomorphisms do however occur. While most single populations will possess two or more alleles in a genetic system, some populations will possess only one of these genes. For example, most populations are polymorphic for the capacity to taste phenylthiocarbamide with both 'tasters' and 'non-tasters' in the population, but in some populations only 'tasters' are to be found. In cases like this it would appear that usually if not invariably only one of the forms is ever monomorphic. In the 'taster' example, there are no populations who consist exclusively of non-tasters. This phenomenon is rarely commented on but has important evolutionary implications, and also acts to reduce estimates of between population variability.

Another striking feature of geographical variation is that it tends to be *clinal*: that is to say that changes in gene frequency tend to occur gradually and evenly over distance. A well known example is the slow decrease in the gene for blood group B in the ABO blood group system as one moves through Europe from east to west; but there are innumerable other cases over both long and short geographical distances.

Because of the widespread existence of clines, putting people into taxonomic categories such as named races seems arbitrary if not artificial. Frank Livingstone of Michigan summed it up some years ago by saying 'There are no races: only clines'. However, the existence of races, and the value of racial classifications are two different things and, more to the point, not all geographical variation is clinal. There are often sharp discontinuities with considerable change in gene frequency in short geographical distances, particularly where there exist barriers to human movement such as mountains and seas.

Perhaps more troublesome from the classificatory point of view, is the fact that different genetic systems tend to have different patterns of geographical variation. Thus, for example, the ABO and Rhesus blood groups systems show different patterns from each other and from the distribution of haemoglobin variants and PTC tasting. If the patterns were totally discordant then classifications would clearly be a nonsense but this is not the case and though no identical distributions exist, many systems show broadly concordant patterns. Because of this one can assign with considerable geographical precision the provenance of a collection of blood samples tested for a number of different genetic systems. One can even make a reasonable guess from a single blood sample, of the continent or major geographical zone in which the blood was collected if it is tested for all the variant genetic systems known. Ironically, this is just about as well as one can do using the characters of old-fashioned physical anthropology such as skin colour, hair form, body physique and facial features like those considered by Huxley and Haddon. Although the genetic basis of these characters has remained so far intractable to analysis such a basis exists and it involves many genes. It is the sampling of many genetic systems which is the most important requirement for assigning geographical provenance.

A long and major interest of anthropologists has been in discovering the evolutionary relationships between different human populations, especially those that are geographically apart and genetically most distinct. 'Evolutionary relationship' has many components and facets but most concern has been with genealogical relationship (or cladistic affinity) particularly in terms of identifying common ancestry and lengths of time since such ancestry. Once a population divides into two (or more) and the parts are isolated from one another they clearly progressively lose relationship with one another in the genealogical sense. Once separated they evolve independently, accumulate their own new genes, and progressively become more genetically different (in the absence of convergence). It has become axiomatic that the degree of genetic difference is a measure of the length of separation and therefore of cladistic affinity. Here it needs to be remembered that as isolation leads to a loss of common ancestry, migration between populations leads to the development of common ancestry, and, evidently, migration and the gene flow it implies will tend to make populations more genetically similar.

Many different human 'phylogenetic trees' have been produced by geneticists and anthropologists using different multivariate procedures and different kinds of biological data. They tend to show many common features, with neighbouring groups being closely related and geographically widely separated groups being most distantly related. From what has already been said it is not surprising that schemes based upon morphological characters are very similar to those based on blood groups and the like both in local studies and global ones. The one area where identifiable genes have clarified an uncertain position is in South East Asia and Melanesia. Although some peoples living in this area have superficial similarities to Africans, with dark skin, broad noses and curly hair etc. blood characteristics align them very clearly with other Asiatic populations. Here we appear to have a case of evolutionary convergence in some morphological traits.

Whilst there is broad agreement about the affinities of at least most living human populations there has been a great deal of controversy about the origin of our species *Homo sapiens* and the timing of the beginning of racial variation in it. Broadly, theories can be categorised as monophyletic or polyphyletic. An extreme form of the latter, put forward by C. S. Coon, (1963) involved various separate ancestries for *Homo sapiens* from different races of *Homo erectus*. Coon was impressed by the fact that a number of geographical variations of the skeleton in *erectus* are to be found in *sapiens* living in the same geographical area. Peking man, for instance, has shovel-shaped incisor teeth as do present day mongoloid groups of eastern Asia. Coon was severely criticised for this theory, particularly as he proposed that different groups crossed the erectus–sapiens divide at different geological times. Because of this the theory was labelled 'racist'. It is, of course, very improbable that any species would be evolved more than once but in this respect it needs to be remembered that the palaeospecies can be a very arbitrary category. If in a single line of descent it is decided that two species exist, then at some level a father would have to be put in a separate species from his son if the fossil record were absolutely complete! It is certainly possible that many different if not all *Homo erectus* populations contributed to *Homo sapiens*. One can well envisage common selection pressures acting to transform one species into the other, particularly since many of the ways in which *Homo sapiens* is different from *Homo erectus* such as brain size and cranial morphology represent 'general improvements'. A 'gradal' change as Huxley would have termed it. The parallel change that would arise from similar selection over many populations could be held within the confines of a single biospecies by gene flow between populations. Such gene flow would not only maintain the integrity of species but also ensure that 'general improvement' genes arising in one group could be transmitted to all and thus favour the parallel change. I like many others long thought that the local exchanges between neighbouring populations through exogamy would provide a network of gene flow which would be sufficient to ensure this pattern, but some simple modelling shows that genes flow much too slowly under such a system. Some long-range migration is a prerequisite but this is not incompatible with the fossil and archaeological record (Harrison, 1984). In a scheme of this kind, racial variation, i.e. genetic variation distinctive of a particular geographical region, can well predate the origin of *Homo sapiens* (Hiorns and Harrison, 1977). All that is required is that local selection forces favour some genes over a long period of time, while universal selection forces are transforming the genetic systems which distinguish *Homo erectus* and *Homo sapiens*.

However, recent molecular evidence particularly from restriction sites in the ß-globin gene cluster (Wainscoat *et al.*, 1986), and mitochondrial DNA genes (Cann *et al.*, 1987) favour a single, monophyletic origin of *Homo sapiens* in Africa and a subsequent rapid spread throughout the rest of the world during which racial variation developed. Mitochondrial DNA is perhaps especially interesting since it appears to evolve much more rapidly than nuclear DNA and is inherited solely through the female line. Cann *et al.* (1987) have hypothesised a true African 'Eve' from about 250 000 years ago. All evolutionary modelling conceptualises an ultimate ancestor or pair of ancestors, but in reality, a single little differentiated population, such as has been through a bottleneck will serve as well, so a 'single Eve' is not a prerequisite. However, the fact that there appears to be much more molecular diversity within Africa (between its various populations) than within other continental areas does suggest a greater antiquity for African *Homo sapiens*. And the additional fact that molecular characteristics of African groups are more diverse from other continental groups than the latter are from each other suggests that non-Africans have a relatively recent common ancestry: possibly from a single group migrating out of Africa. Fossil evidence also indicates the presence in Africa of forms skeletally similar to modern man earlier than elsewhere, with dates of at least 100 000 years and perhaps earlier.

Nevertheless the molecular evidence still needs to be treated with some caution, especially as it is necessarily based on small samples. It indicates, for example, that San groups separated from other human lineages 220 000 years ago and that the division between Asians and Europeans was as recent as 5500 years ago (Barton and Jones, 1983). These are clearly nonsense figures and suggest that mitochondrial DNA is not evolving at the same rate in all human populations.

The monophyletic origin hypothesis raises another problem. What happened to all the non-African erectus populations? Were they all brought to total extinction as the descendants of one small group of Africans (themselves the scions of an African *Homo erectus*), very rapidly increased in number and spread over the whole of the rest of earth? It seems intrinsically very unlikely, especially as mitochondrial genes are able to cross the boundaries between biospecies more easily than nuclear genes (Barton and Jones, 1983). Clearly the crucial question in the debate is the one of geological timing of events. Everyone accepts that the first stages of hominid differentiation were in Africa and that *Homo erectus* colonised the Old World from there. But whether or not there was a repeat process by *Homo sapiens* must remain for the time being conjectural.

The final question to be considered about human geographical variation is its causes. This is the most controversial issue of all, yet also the most fundamental. On its answer depends all the issues so far addressed and particularly those concerning the reliability of the phylogenetic reconstructions we have just considered.

Broadly considered there are two kinds of explanation for genetic differences within and between populations. Both recognise mutation as the ultimate source of all new genetic variation but in one view most variants are thought to be inconsequential to living, for example adaptively neutral, whilst in the other view they are thought to affect, directly or indirectly, sooner or later, survival and reproduction, i.e. Darwinian fitness. If genes are neutral, or near neutral, then changes in their frequency over time occur by the chance factors – or the stochastic processes – of genetic drift, founder principle and bottlenecks. If on the other hand genes are varyingly adaptive, then evolutionary change in a population occurs, essentially, through natural selection. It is, of course, the operation of evolutionary forces within populations which produces differences between populations.

There are many varying consequences from these two theories. In the neutral theory of evolution, so persuasively developed by Kimura (1983), evolutionary change occurs at more or less constant rate, at least in any one genetic system, and the ultimate determiner of that rate is the mutation rate in the system. Populations genetically isolated from one another, therefore diverge from one another at constant rate, and levels of genetic similarity between populations is an accurate reflection of their genealogical (cladistic) relationship, i.e. the time since they separated. If, however, natural selection is the main force operating on genetic variants, then evolutionary rates can be very variable. Natural selection operates as often to hold systems constant as to produce change, and evolutionary rates are essentially controlled by environmental factors; it is they that determine selection pressures. Evolution can in a single lineage at some time be rapid, at others slow and at yet others have practically stopped. There is therefore no necessary relationship between the level of genetic similarity between populations and time to their common ancestry. Two very different groups may have had a very recent ancestry, especially if they have lived in very different environments, whilst two similar groups may have a very distant ancestry, especially if they have lived in similar environments. Convergence of populations is also likely if they have been exposed to similar selection pressures.

All biologists recognise that natural selection has played a central role in evolution and evidence of environmental adaptation is ubiquitous. Most evolutionists also recognise that stochastic processes have been important, particularly over short periods of evolutionary time, and especially in the non-coding sequences of the DNA. Notwithstanding, there are clearly 'two camps' when it comes to interpreting the causes for genetic variation the 'neutralists' and the 'selectionists'.

The 'neutralist' view, now the predominant one, is based upon various independent kinds of evidence. First, it does appear that each kind of protein molecule evolves at a more or less constant rate, when considered on the macro-geological time scale covering the differentiation of species, genera, families etc. Secondly it would seem that many amino-acid substitutions in a protein do not and indeed could not affect its function: many segments appear to do no more than join together important functional sites. Then, of course, because the genetic code is redundant numerous base changes are not reflected in protein structure at all, and intervening sequences in the DNA have no product. Thirdly neutralists, following arguments of J. B. S. Haldane, have pointed out that were every polymorphic genetic locus to have a fitness variable attached to it, the genetic load imposed over all the loci would be so great as to cause extinction, since every individual would be bound to carry more than enough disadvantageous genes to kill him! Finally, neutralists point out that apart from a few spectacular cases, mainly concerning haemoglobin variation, there is little actual evidence for natural selection operating on human biochemical variety – or molecular variety in any organism for that matter.

There is, of course, a selectionist retort to all these points. One might well expect variable evolutionary rates to 'even out' over substantial geological time as organisms pass through various cycles of environmental change. And it may not be irrelevant that different kinds of protein evolve at very different rates. Then, so far as function is concerned, this needs to be assessed *in vivo*; and perhaps, more importantly, it is only one aspect of the effect of a genetic change, which ought to be evaluated, were it possible, within the total 'economy' of development and homeostasis.

The problem over 'genetic load' only arises when one fits absolute fitness values to different genotypes. If fitness varies according to what other genotypes are present then any difficulty disappears, and we usually consider Darwinian fitness in relative terms. It has also been argued that selection might be expected to operate over whole arrays of genes, rather than individual loci, since the ultimate unit of selection is the totality of an individual's phenotype.

This may be a reason why we know of so few examples of natural selection operating, but even detecting quite strong selection on individual gene loci is very difficult, it requires much larger samples than have usually been collected in 'field' situations. It can hardly be coincidental that the strongest evidence for selection in man comes from genetic systems which have been longest known and most studied, e.g. ABO and Rh blood groups, haemoglobinopathies, Glucose-6-Phospate Dehydrogenase deficiency, secretor status and PTC tasting status.

Even 'neutralists' acknowledge that the quantitative variability shown in morphological and physiological traits, like body form, skin colour and blood pressure has selective significance. If this is so, it seems intrinsically unlikely that the variation at the biochemical and molecular level would be neutral, at least some of the latter must represent the polygenes of the former!

These considerations are vital in considering the patterns of human geographical genetic variation which were described earlier. Neutralists see recency of common ancestry as the overwhelming determinant of these patterns. Here it needs remembering that whilst separated populations progressively lose ancestry with time, populations can gain common ancestry through intermixture. Clines are perceived in the neutralist view as the consequence of populations diverging through drift but held together to form a pattern of changing gene frequencies through intermixture, with the greatest intermixture occurring of course between neighbouring populations.

Although ancestry is clearly acknowledged as an important cause for genetic similarity between populations in the selectionist view, especially in the short term, on this model the ultimate determinant of the genetic characteristics of a population is the environment in which the population exists and the selection pressures that environment imposes. Populations in similar environments will become genetically similar irrespective of their ancestry. While clines may be the result of gene flow they can also arise through a gradient in selection pressures, or through a mixture of selection and gene flow. And discordance in character distribution comes about because the geographical distribution of selective forces for different characters is not the same. Thus while sickle-cell haemoglobin, thalassaemias and hereditary persistence of fetal haemoglobin have been essentially confined to regions of endemic Falciparum malaria, skin colour variation is concordant with the geographical distribution of ultraviolet radiation and nasal index with the distribution of vapour pressure and other measures of the evaporative power of the air.

As with many controversies the reality in the selectionist/neutralist debate probably lies somewhere near the middle of the polar views, with chance playing more of a role than selectionists admit, and selection being more important than the neutralists acknowledge. It is quite possible that genes once adaptive become neutral; indeed one might expect such changes to occur in the constant co-evolution of hosts and parasites. It is also possible for originally neutral genes to become adaptive; there is good evidence for duplicated genes taking on new functions. It is also worth remembering that while selection is usually treated as a deterministic force, there are many stochastic elements in its action.

Whatever the ultimate answer, there is a quite different factor which has certainly played an overwhelming influence in determining the genetic composition of the present-day human species: comparative population growth. Throughout most of human evolution, hominids existed as hunter-gatherers organised into small local populations. It is in such settings that genetic differentiation of populations largely occurred. Since the neolithic, however, certain of these groups have grown dramatically in size, and in the process assimilated and genetically swamped, or brought to extinction most of the others. The process is well documented in the recent expansion of European populations, particularly in the New World and Australasia. But this was by no means a unique phenomenon and there is plenty of archaeological and historical evidence for earlier expansions in Asia and Africa. It is these growths and expansions which have most profoundly affected genetic distributions and, as F. S. Hulse has so cogently pointed out, which populations grew and expanded was dictated solely by cultural and historical factors, and had nothing to do with most if not all their genes.

References

Barton, N. and Jones, J. S. (1983) 'Mitochondrial DNA: New Clues about Evolution', *Nature*, vol. 306, pp. 317–318.

Cann, R. L., Stoneking, M. and Wilson, A. C. (1987) 'Mitochondrial DNA and Human Evolution', *Nature*, vol. 325, pp. 31–6.

Coon, C. S. (1963) *The Origin of Races* (London: Jonathan Cape).

Harris, H. (1980) *The Principles of Human Biochemical Genetics* (3rd edn) (Amsterdam: Elsevier, North-Holland).

Harrison, G. A. (1984) 'Migration and Population Affinities', in A. J. Boyce (ed.), *Migration and Mobility* (London: Taylor and Francis).

Hiorns, R. W. and Harrison, G. A. (1977) 'The Combined Effects of Selection and Migration in Human Evolution', *Man*, vol. 12, pp. 438–45.

Huxley, J. S. and Haddon, A. C. (1933) *We Europeans* (London: Jonathan Cape).

Kimura, M. (1983) *The Neutral Theory of Molecular Evolution* (Cambridge: Cambridge University Press).

Lewontin, R. (1982) *Human Diversity* (New York: Scientific American).

Wainscoat, J. S., Hill, A. V. S, Boyce, A. J., Flint, J., Hernandez, M., Thein, S. L., Old, J. M., Lynch, J. R., Falusie, A. G., Weatherall, D. J. and Clegg, J. B. (1986) 'Evolutionary Relationships of Human Populations from an Analysis of Nuclear DNA Polymorphisms', *Nature*, vol. 319, pp. 491–3.

11 Evolution and Ethics

Patrick Bateson, FRS

INTRODUCTION

In 1893, Thomas Henry Huxley delivered the Romanes lecture in Oxford, giving it the title 'Evolution and Ethics'. It was a polished performance, highly literate, confident and progressive. He established a clear distinction between savage, selfish behaviour generated by biological evolution (or the 'cosmic process' as he called it) and civilised, altruistic values resulting from the 'ethical process'. The lecture ended on an optimistic note:

> I see no limit to the extent to which intelligence and will, guided by sound principles of investigation, and organized in common effort, may modify the conditions of existence. . . . And much may be done to change the nature of man himself.

Who could doubt it? The best aspects of Victorian thought were themselves the end-products of the evolution of morality and would be harnessed to bring the rest of the world up to the same high standard.

Fifty years later his grandson Julian stood at the same podium and fittingly gave his own, rather more critical view of what he called 'Evolutionary Ethics'. Julian Huxley's own Romanes lecture expressed the same humanity and liberalism found in his grandfather's, but its style was quite different. It was dense, reflective and somewhat hesitant. Julian was more conscious than Thomas Henry of how relative are moral judgements and how much they differ between cultures and even between people within a culture. Consider, for instance, differences in attitudes to the mother and the unborn child in relation to abortion. He wrestled with the conflicts which can arise when different ethical judgements point in opposite directions. He was also intensely aware of how the apparently unlearned characteristics of morality are the product of an individual's own upbringing. Despite the problems which he had clearly perceived, Julian would have liked to explain the historical development of socially transmit-

168

ted morality (the 'psychosocial process') in the same general terms as Darwinian evolution. Both processes, he argued, generated behaviour patterns that *worked* better than rival alternatives in the conditions in which they were found. However, the very same processes that can generate a good fit between behaviour and the physical environment can also lead to runaway change when the behaviour involves choice between or control of members of the same species (Bateson, 1988b). The most familiar biological example involving genetic inheritance is 'sexual selection' where mate choice can lead to the evolution of bizarre ornaments such as the massive tail of the peacock. These evolve because they win mates, but they are not adaptations to problems set by the physical environment. Indeed, they can reduce the chances of survival. It is not difficult to see how non-adaptive preferences derived from copying others could have arisen in this way. For instance, if people adopt the standards of dress of the majority, but then express a preference for clothes that are slightly different, historical movement in majority habits can and, of course, does take place (see Boyd and Richerson, 1985 for a formal model involving social inheritance).

In his introduction to the book in which the two Huxley lectures were reprinted, Julian posed the questions that seemed to him critical.

Is there any external standard for morals? Any touchstone by which goodness may be recognised, any yardstick by which it may be measured? Does there exist any natural foundation on which human superstructure of right and wrong may safely rest, any cosmic sanction for ethics? (Huxley and Huxley, 1947)

I suspect that he really wanted to answer: 'Yes, yes, yes.' Part of him deeply believed that both Darwinian evolution and the 'psychosocial process' had the capacity to uncover goodness. If we could only understand what these processes had done, then we should know how to conduct our lives. In this mood, if he did not get sucked right into it, he sailed perilously close to the whirlpool called the 'naturalistic fallacy'. Given what is known about human behaviour, this is the way we *ought* to behave. As Williams (1983) has noted, there is a vast and important difference between such a justification for human morality and claims of a very different kind, namely, given what is known about human behaviour, we *cannot* behave in any other way. Or, more modestly, given what is known, we are likely to have developed a particular form of morality in a particular set of conditions. This last claim is properly seen as historical explanation, rather than as justification. It is an issue which I shall return to later.

The relativism which Julian Huxley correctly detected in ethical judgements and the sheer complexity of cultural evolution deterred biologists from contributing much more to the debate about the origin of ethics for the next 30 years. In the 1970s, however, the growing interest in the Darwinian evolution of cooperation led to a fresh attempt by biologists to sail around the naturalistic whirlpool. E. O. Wilson (1978) and Alexander (1980) in particular, tried to explain the conduct of human affairs in biological terms. The moral rules which guide us as to how we should behave towards others seem to be partly rooted in a spirit of cooperation. If it was possible to provide a biological explanation for cooperation, then maybe it would be possible to provide a biological explanation for ethics – or so it seemed to the people who were captivated by this prospect. So, why did animals evolve ways that benefited fellow members of their own species? Three evolutionary explanations have been proposed for non-manipulative social cooperation: (a) The individuals are closely related; (b) the individuals mutually benefit; and (c) the surviving character is a property of many individuals. These explanations have been persuasive (admittedly to varying degrees) and have led to the fresh attempts to explain consciously intended altruism in the same terms.

Some philosophers have responded enthusiastically to this renewed interest by sociobiologists in the old problem of where ethics come from (e.g. Ruse, 1986). However, the sociobiologists' efforts have not been greeted with universal enthusiasm (e.g. Kitcher, 1985). Indeed, I doubt whether Julian Huxley would have much liked some of their stronger claims. Even so, the sociobiology controversy has brought back Huxley's interests in the origins of ethics to the centre of the academic stage. Furthermore, the debate has stimulated increasingly sophisticated discussion about the interplay of genetic and social inheritance (e.g. Boyd and Richerson, 1985).

In a short essay I cannot do justice to the richness of modern thought about the ways in which Darwinian evolution, relying largely on genetic inheritance, and cultural evolution, relying largely on social inheritance, relate to each other. Instead, I shall try to sketch elements which I regard as crucial in the historical process and then show how the interplay between these elements may have been involved in the historical development of a particular instance of human morality. The elements are cooperation, conformism and culture and the particular example is the incest taboo. I shall begin by considering the three ways in which cooperation might have evolved by Darwinian evolution.

CHARACTERS SHARED BY KIN

The first explanation for the evolution of cooperation which I shall consider has been the special domain of sociobiology. The idea of 'kin selection' is an extension of the intuitively obvious point that animals will often put themselves at risk and do things that are bad for their health in the production and care of their offspring. The use of the term 'altruism' in sociobiological discussions was unfortunate because of its moral connotations. By begging the crucial question of whether or not ethics grew out of cooperation, the more fundamental principle of how cooperation evolved was shrouded in controversy. Also, as Williams (1983) noted, the term 'altruism' clearly implies a conscious intention to help another, whereas the mere performance of an action that increases the chances that another will survive and reproduce does not.

Leaving aside the controversy over the use of altruism, the explanation for the evolution of cooperation is this. If the beneficiary of the action is sufficiently closely related to the performer and if the genetic difference between the individuals that perform such actions and those that do not is also small, then in the course of evolution individuals helping collateral kin are likely to become increasingly frequent in the population.

The evolutionary principle can be perceived more clearly, perhaps, when a non-behavioural example is used. Consider those insects, like wasps, that are conspicuously marked and are unpalatable to their predators. Birds that eats wasps are unlikely to repeat the experience, since birds learn quickly. This does not help the wasps that died. However, in the ancestral condition, the few wasps that were conspicuously marked were likely to be closely related since nest mates are usually sisters. Those that died provided protection for those that survived by making them less prone to predation. As a consequence, conspicuous yellow and black abdomens may have spread until all wasps were conspicuously marked in the same way. It is not difficult to see how a precisely similar argument can be mounted for *care* directed towards close relations. The point is that the giving of aid to a relative may evolve simply because the expression of that character increases the probability that its expression will recur in later generations (Hamilton, 1964).

MUTUALISM

The second explanation for cooperation is sometimes known as 'mutualism within a species' (e.g. West-Eberhard, 1975). Two cooperating individuals are not necessarily related, but they are both more likely to survive and reproduce themselves if they help each other. This category includes the favourite example of the games theorists, the Prisoner's Dilemma. If one prisoner helps the police by incriminating his friend, he gets a lighter sentence – unless his friend has done the same to him, in which case they both get heavy sentences. If neither cheats, they might both go free. What should they do? Axelrod (1984) staged a competition for the theorists to discover who could find the best set of rules for the prisoners. In repeated encounters, the solution that consistently produced the highest pay-off (i.e. shortest run of prison sentences) was TIT-FOR-TAT. For a specified set of pay-offs, both prisoners' selfish interests are best served by cooperating with the other at the outset. Only if one of the prisoners cheats does the other retaliate on the next occasion that he is able to do so. After one retaliation he returns to cooperation on future occasions. Admittedly, this example is somewhat contrived, so I shall focus on a straightforward piece of biology, the joint care of their offspring by both sexes.

Even when both parents care for young, their interests do not coincide. They certainly have a common interest in their offspring's survival, but they have diverging interests inasmuch as each one might be able to increase its own reproductive success by spending time seeking extra mates elsewhere. In many species of birds, in which both sexes normally care for the young and one parent dies or disappears, the remaining mate increases the time and energy it devotes to caring for the young. This frequently observed event raises the question of the extent to which an animal can be a 'free-rider' on the efforts of its mate.

Free-riders, who leave all parental care to their mates, will not evolve if the respective amounts of care given by cooperating parents maximise the reproductive successes of each parent in those conditions (Chase, 1980). Each animal involved in the cooperative care of young has an independent set of conditional rules about what to do if the help provided by its partner changes. These may not be the same for both sexes and will depend on the opportunities available for getting other matings. The rules will be the product of Darwinian evolution in the sense that the animals that had most offspring in the past would be those that most nearly found the optimum for a particular set of conditions.

Houston and Davies (1985) have provided an illustration of how such postulated rules might work in a common garden bird, the Dunnock. Dunnocks were believed to be monogamous, but only some have a stable relationship between one male and one female. Some are polygynous, some are polyandrous and, even more remarkably, some breed in combinations of several males and several females (see Davies, 1985). In all breeding arrangements the amount of effort put into feeding the young increases with the number of young. Taking that into account, in monogamous pairs the female is responsible for slightly more than half of feeding. However, the female reduces the number of feeds to the brood when she is helped by two males. Houston and Davies (1985) found that the female's feeding rate is about 7 per cent less when she has two mates. She does reduce her own rate of feeding the young when she has more help, but she certainly does not give up altogether – as might be naively expected if she operated on the principle of unenlightened self-interest.

Another example of mutualism is reciprocated aid (see Trivers, 1985). I do something for you today in the expectation that you will return the favour tomorrow. It is thought that such behaviour may have evolved in more complex animals. However, the theory of repeated interactions leading to reciprocated cooperation is not well developed for large groups. Taylor (1976) suggests that it can work but that the reciprocal arrangements are complex and readily susceptible to disruption. These problems become more acute as the group size gets larger. Since people do cooperate in large groups, it would probably be unwise to pin all our hopes of explaining human behaviour on this type of mutualism.

COLLECTIVELY GENERATED CHARACTERS

The third evolutionary explanation for cooperation, commonly referred to as 'group selection', is the most controversial, largely because a good argument has been confused with a bad one. The bad argument is that animals ought to be nice to each other for the good of the species. The idea is inadequate because any individual that breaks the rule and behaves in a way that benefits itself at the expense of other members of its species will eventually populate the world with individuals that behave in the same self-serving way. The good argument is that some assemblages of individuals may, through their concerted efforts, generate an outcome that puts their group at an advantage over other groups. In other words the character, which the metaphorical hand has supposedly selected in the course of Darwinian evolution, may be formed by more than one individual. The characteristics of the whole entity provide the adaptations to the environment. One assemblage of individuals, acting as an organised system can compete with another in the strict Darwinian sense of differential survival.

The possibility of group characters changing in Darwinian fashion is not in question among serious evolutionary biologists. However, the consensus in the last ten years has been that the conditions for such evolution were too stringent, since groups are usually much slower to die off than individuals and individuals can readily move from one group to another (see Maynard Smith, 1976). The consensus was probably formed too readily and has been under attack in recent years (see Wade, 1978; Wilson, 1980).

The essential point is that the outcome of the joint action of individuals could have become a character in its own right. The nature of the argument may be perceived most clearly in the arrangements of different species that are obliged to live together in symbiotic partnership. A good example is provided by the lichens. While they look like single organisms, lichens are composed of algae and fungi fused together in obligatory partnership. In Darwinian terms, though, overall features of a lichen might enable it survive better in a given environment than a lichen with other characteristics. Even though the character is replicated in an 'offspring' lichen by the independent reproduction of the component algae and fungae, the mechanism of inheritance is irrelevant to the evolutionary process. So long as offspring characteristics are correlated with parental characteristics, it does not matter how they got like that.

To take a specific example, suppose that in one 'individual' lichen, an algal mutation and a fungal mutation have given rise to a product that makes the lichen less tasty to reindeer (which are lichen specialists). The less palatable lichens will survive better than those without the mutations. This is not because of competition between algae or between fungae, but because of the combined effects of the mutations on the entity of which they are a part.

The importance of this argument to the evolution of cooperation is that, if the conditions were right, the outcome of the joint actions of individuals in social groups would have changed as the result of Darwinian evolution. It is important to appreciate that this perfectly straightforward Darwinian argument does not undermine what we know about genetics or return to muddled good-for-the-species thinking. It merely draws attention to a higher level of adaptation. This requires acceptance that the characteristics of social groups are the emergent properties of the participating members and the general logic of Darwinian theory applies as much to these characters as it does to those of individual organisms.

The emphasis that I give to this last evolutionary explanation for human cooperation has been encouraged by an argument of Boyd and Richerson (1985) in developing their general thesis about the interplay of genetic and social inheritance. They argue that conformism can amplify differences in culture between groups and, with a predisposition to imitate the most common form of behaviour in the group, differential survival of populations may follow. Under these conditions group selection can work even when groups are large, extinction rates of individuals are higher than those of the groups in which they live, and migration rates between groups are substantial. Since the evolution of cooperation might depend so critically on conformism, we should look more closely at this aspect of behaviour.

It makes biological sense that if a member of a social group behaves aberrantly, he or she should be tested and if necessary checked. The aberrant pattern of behaviour might have various undesirable consequences for the group members. It might draw the attention of predators or unwelcome competitors. It might disrupt communal activities on which the survival and reproductive success of the others depended. It might indicate disease in the aberrant individual that was potentially dangerous to the others. Whatever the explanation, the attacking of aberrant individuals is often seen by people studying social animals under natural conditions. For instance, I have seen an Ivory Gull that had been injured and was flapping about on the ground being swooped upon and pecked by the other members of the colony in which it lived.

The conformism of humans is obvious enough and, indeed, is much studied by social psychologists (reviewed by Boyd and Richerson,

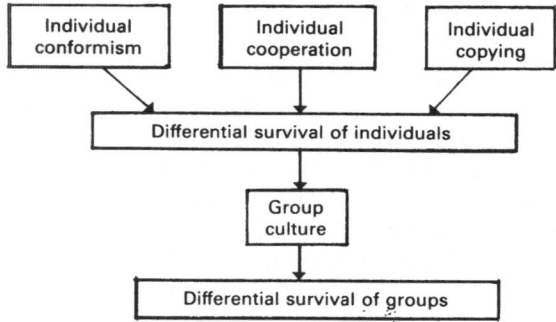

Figure 11.1 The generation of group cultures. Predispositions to conform, to cooperate and to copy others are separately beneficial to the individuals that express them and thereby evolve. When all three characters are present, group cultures are generated. The collectively generated groups become beneficial and thereby evolve.

1985). When individuals voluntarily pass at least some of the control of their behaviour to another person, we call what has happened to them 'hypnotism'. However, rather than regard this simply as a stage trick or some peculiar epiphenomenon of human psychology, it is more realistic to treat hypnotism as an extreme manifestation of normal social behaviour (Zangwill, 1987). All of us all the time readily hand over the running of much of our lives to somebody else. We obediently have our actions planned for us when we are at work and equally willingly do as we are told on package holiday tours. This acceptance of external authority is one highly efficient way in which to live cooperatively. It also means that whole societies can rapidly come to share the same patterns of socially transmitted behaviour.

In summary, then, the predispositions to cooperate, conform and copy the behaviour of others may well have created the conditions that enabled Darwinian evolution to operate on the characters of groups (see Figure 11.1). Furthermore, if some groups survived better than others by virtue of these predispositions, then cooperation, conformism and copying would have been strengthened. Such positive feed-back would have meant that the evolutionary change in the character of these predispositions would have been explosive. Indeed, as we learn more about human evolution, exponential change appears to be one of its dominant characteristics (see Foley, 1987).

HUMAN COOPERATION AND HUMAN ETHICS

The arguments about the evolution of cooperation in animals have persuaded most biologists that such behaviour can be readily understood in terms of Darwinian theory. People differ, of course, in how much emphasis they wish to place on each of the explanations. My own view is that kin selection, advantages to the individual of mutualism, and the differential survival of characters collectively generated by more than one individual, were all involved in human evolution (Bateson, 1988a). However, explaining the evolution of cooperative behaviour is not the same as explaining the emergence of ethical behaviour. Explanations for differences in the cooperative behaviour of animals in terms of the Darwinian theory of evolution, rest on an assumption that the behavioural differences depend on genetic differences. Admittedly, within a species, differences may well depend on environmental conditions, but the conditional rule which generates the variation is usually treated as a character that has itself been subject to Darwinian evolution and is now shared by all members of the species. These assumptions are rarely made in the case of differences in ethical behaviour of humans. Indeed, this seems to show a common sense reluctance to push a biological argument too far. To take the abortion issue again, it would be far-fetched to argue that the difference between concern for the mother and concern for her unborn child represents the expression of an unlearned conditional rule possessed by everybody.

A possibility for muddle needs to located, however. Confusion arises if a behaviour pattern, which was brought about by rational thought, also serves individuals well in terms of increasing their chances of survival and their reproductive success. If the rules for such cognitive processes arose in the course of Darwinian evolution, then it is sometimes argued that so too did the behaviour pattern (e.g. Wuketits, 1984). In the early nineteenth century, William Paley argued that just as the obvious function and workings of a watch imply conscious design, so do those of living organisms (see Dawkins, 1986). In modern times, the argument has been stood on its head. If a behaviour pattern resulting from rational thought looks well designed, then it is the product of the blind historical processes that gave rise to the capacity for rational thought. This mistake, as I believe it to be, is exposed in those cases in which human cleverness is turned towards maladaptive practices such as the procuring of addictive drugs.

It makes a great deal of sense to suppose that the complex processes of the brain that generate behaviour were themselves the result of Darwinian evolution. Individuals that had such capacity survived better than those that were less well endowed. It makes no sense at all, however, to argue that the present-day *products* of human brain power are generated by Darwinian evolution when they increase the chances of survival and are not generated by natural selection when they have the opposite effect. The difficulty, about which we need to be clear, is that good design may arise in a number of different ways. Generating variation, differential survival and onward transmission of the surviving character can apply to the production of ideas in the head of an individual or to the production of values within a culture. However, the formal similarities with Darwinian evolution should not encourage confusion of the three processes, all of which involve 'selection' in the sense that the version that works best is the one that survives.

Despite the evident difficulties, what should we make of cases where behaviour patterns resulting from Darwinian evolution look very similar to the products of Huxley's psychosocial process? Predispositions to cooperate and morally guided intentions to help others look alike in their consequences. None the less, we need to think very carefully about how one might turn into the other. Williams (1983) describes the pitfalls in simply focusing on the superficially similar outcomes. When we are clear about the differences in the historical processes, we need to examine how, in particular cases, the outcome of one might have impinged on the other. Specifically, how do genetically transmitted predispositions influence socially transmitted ethical values? In the next section I shall examine inbreeding avoidance and the incest taboo in order to see how far we might be able to go in providing an answer to such a question.

ORIGINS OF INCEST TABOOS

Incest taboos highlight both the problems and the possibilities of using biological knowledge to explain the origins of human morality. The question is whether evolutionary thinking can add anything to our understanding of social prescriptions about mates. Many commentators have thought it was enough to point out that incest taboos exist because they generally have the effect of preventing inbreeding. Since inbreeding has biological costs and incest taboos reduce the chances of incurring them, it was supposed that taboos must be a product of Darwinian evolution. They function to increase reproductive success. The supposed regularities of human morality were attributed to the workings of adaptive rules, so providing an evolutionary explanation for part of human culture. The conclusion is too facile. First, the advantages of stopping other people from incurring the consequences of their own folly are not immediately obvious. Second, the argument does not explain the marked variation between cultures in which categories of relations are prohibited from having sexual relations. Finally, and most seriously, it is not at all clear, as Williams (1978) has pointed out, how a genetically transmitted inhibition was somehow converted into a socially transmitted prohibition.

Two quite distinct arguments have been mounted to circumvent these objections and explain the origin of incest taboos (see Figure 11.2). Since the arguments are sometimes confusingly conflated, it is helpful to keep them separate even though they are not mutually exclusive. The first is the one considered by Williams (1983) and runs as follows. Humans, being observant and intelligent, spot the consequences of matings between close relatives and resort to rational collective action. In effect, they make safety laws. The incest taboo is equivalent to a legal requirement to wear seat belts or crash helmets.

The second argument is derived from Westermarck (1891). He suggested that satisfying sexual relationships are not formed between people who have spent their childhood together. Prohibitions may then have arisen from the social pressure directed against particular types of unorthodox behaviour. This is because people often strongly disapprove of others who behave in unusual ways. The most obvious example is the moral repugnance that many people show for homosexuality between consenting adults. Similarly, social disapproval has commonly been directed against left-handers.

I have argued elsewhere that the safety law argument is not especially plausible because the genetic costs of intermittent inbreeding are not easily observed. A universal taboo requires universal knowledge of the ill-effects of inbreeding (Bateson, 1983). I liked the Westermarck idea because it was supported by both animal and human evidence. Humans probably resemble birds and mammals in preferring as mates individuals who are bit different but not too different from the members of the opposite sex whom they knew earlier in their lives. Furthermore, conformism is obvious enough in humans. A potential weakness in the argument is that it leaves

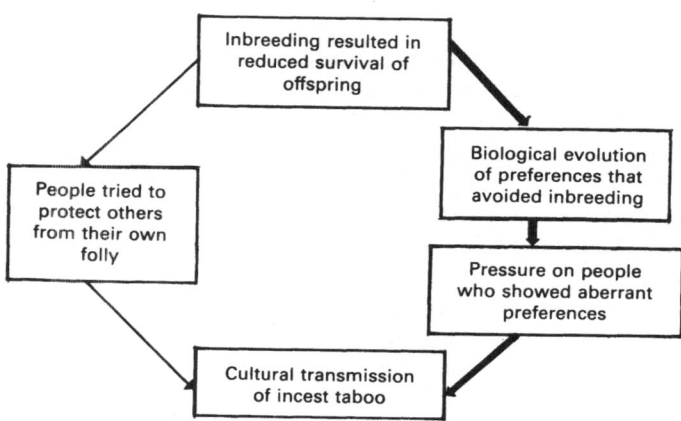

Figure 11.2   Two suggested ways in which incest taboos might have
developed from the biological disadvantages of inbreeding. The first
implies that people noticed the ill-effects of inbreeding and made safety
laws to protect others (and society as a whole) from such consequences.
The second argument proposes the Darwinian evolution led to preferences
for mates that were different from members of the opposite sex who were
familiar from early life (and therefore likely to be closely related). As a
result of social conformism, pressure was brought to bear on people who
expressed sexual preferences for close kin and who therefore behaved
differently from the majority. This second route, which is regarded as
being the more plausible, is indicated by the thicker arrows.

unexplained why social disapproval is directed at some forms of
unusual behaviour and not at others. To avoid circularity it is strictly
necessary to identify independently three predispositions, not two.
These are the predisposition to prefer slight novelty, the predisposition to make others do what you do yourself, and a predisposition to
react particularly strongly to certain types of aberrant behaviour.
None the less, the general form of the argument is attractive because
it suggests how cultural transmission of a set of values may arise as a
consequence of genetically transmitted patterns of behaviour that
presumably originally arose by Darwinian evolution. The conformism generating prohibitions is not an adaptive response that evolved
in the service of maintaining optimal outbreeding. It arose for quite
different reasons, as I have already discussed, and among its other
consequences happened incidentally to amplify the beneficial effects
of the inhibitions.

To summarise the argument, humans may have evolved like many other animals so that experience obtained in early life provided standards for choosing mates when they were adult. Normally the experience was with close kin so that, when freely choosing a mate, a person preferred somebody who was a bit different but not too different from close relations. This system evolved because those who did it avoided the maladaptive costs of both extreme inbreeding and too much outbreeding. Consequently, the system was more likely to be represented in subsequent generations. This step could have occurred long before language evolved.

The next step was that, for other reasons to do with the biological benefits of cohesion, conformity was enforced on all members of the social group. Those who did particular things that the majority would not do themselves were actively discouraged. In sexual affairs, social pressure was put on people who interested themselves either in individuals who were very closely related or who were unfamiliar to members of the group. As language evolved, prescriptions about mating were transmitted verbally to the next generation. In this way taboos and marriage rules characteristic of a culture came into existence.

While prohibitions are broadly correlated with inhibitions, they did not arise directly as the result of biological evolution. A merit of this dissociation is that it then becomes possible to account for some of the variation found in incest taboos, which take many different shapes and forms in different societies. They sometimes include certain types of cousin but not others. Some forbid marriages with people who are familiar but not genetically related in any way. Prohibitions on sexual relations with parents, siblings or children, are nearly always universal – but not quite (see Hopkins, 1980). The crucial link between the predisposition and the moral standard is the way people acquire their preferences. What is seen as a pattern of 'normal behaviour' is itself influenced by the pattern of early experiences that are common to that society. The implication is, therefore, that child-rearing practices and taboos will be correlated. Cultural differences in prohibitions should be related to the categories of persons who are familiar from early life. So if children grow up with some types of cousin but not others, the most familiar cousins should then be prohibited as sexual partners, whereas the others should not. The same goes for non-blood relations.

The neo-Westermarckian hypothesis accounts for some cultural differences, which gives it welcome heuristic power (see Bateson, 1983). Nevertheless, it would be exceedingly difficult to argue that all the variation in human marriage laws could be explained in such terms. This is because the hypothesis is silent about the role of power, property and many other factors that are likely to have been important in the formulation of culturally transmitted marriage rules. Obviously it does not (and should not) exclude them. On the contrary, it alerts us to the ways in which different influences might have impinged on the historical process. For that reason, as much as anything else, I share Kitcher's (1987) belief that judicious application of ideas that were originally rooted in biology will aid the understanding of human behaviour.

In summary, then, I have tried to show how the sense of morality about mating preferences grew dynamically out of other features of human behaviour. The *combination* of capacities was crucial. Biological evolution shaped a predisposition to prefer members of the opposite sex who are a bit different but not too different from people we knew when we were young. Biological evolution also shaped a predisposition for social conformism. The outcome of these predispositions is social pressure directed at those who do not exhibit the same mating preferences as others. When linked to a further human capacity, that of language, the historical results of such social pressure are cultures with verbal rules for sexual conduct (see Figure 11.3). Any given end-point of the confluence was not the direct result of Darwinian evolution. It came about because of other human capacities that were. Morality is a side-effect. The implication is that, as child-rearing practices and other conditions of human social life alter, so eventually does the nature of sexual morality.

CONCLUSION

I have tried to develop some arguments which, I think, grow out of the position adopted by Julian Huxley in his 1943 Romanes lecture. He was properly reluctant to ground his evolutionary ethics on the naive belief that every ethical judgement could be traced back to some pattern of behaviour that was of direct advantage to the individual who expressed it. He sought to bring culture and individual development into the historical explanations for ethics. If one is to be frank about it, though, the contributions of biology in the last half century have not provided any coherent *justification* for morality. After much confused debate, we have been alerted to the fact that superficial similarities between predispositions and ethical standards

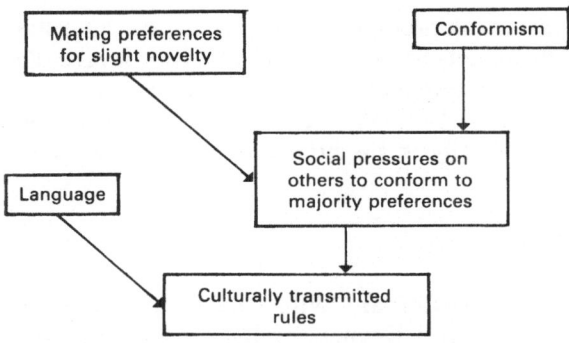

Figure 11.3   The ways in which capacities shaped by biological evolution might have led to the cultural transmission of incest taboos.

do not mean that they are directly linked to each other. Furthermore, adaptive predispositions may lead to the expression of characters with emergent and even non-adaptive features.

If Huxley hoped that one day it would be possible to offer a 'cosmic sanction for ethics', biology has failed to deliver. What it has done is suggest ways in which human predispositions, shaped by Darwinian evolution, have played their part in influencing the historical development of socially transmitted norms of behaviour. I have given one example in this essay. Many other examples are discussed by Hinde (1987).

Modern thought about cultural evolution also suggests that Julian Huxley was mistaken about the adaptiveness of ethics. It no longer looks remotely plausible that every ethical judgement represents a 'good' solution to some present (or past) problem in the social environment (see Williams, 1983). Social transmission of culture may acquire a momentum that is unrelated to external conditions. The combination of conformity and preference for slight novelty found in the expression of culture, can clearly spark off a runaway process. As a consequence, the ethical positions which are adopted need bear no close relationship to what works best in relation to the requirements of the environment. They may represent the cultural equivalent of the peacock's tail.

If he was mistaken in some of his hunches, Julian Huxley foresaw the need to bring the insights derived from evolutionary biology together with those derived from psychology. Here his approach has been strikingly vindicated. What we see now is an attempt to understand historical processes in terms of the interplay between many different influences operating at different levels. The rhetoric that derived from attempts to nail everything down to single causes has come to look intellectually ugly and deeply false. Julian Huxley clearly felt that deeply in his bones and, in that respect, was way ahead of his time.

Acknowledgements

I am very grateful to Bernard Williams for his advice and to him, Robert Hinde and Milo Keynes for their comments on an earlier draft of this chapter.

References

Alexander, R. D. (1980) *Darwinism and Human Affairs* (London: Pitman).

Axelrod, R. (1984) *The Evolution of Cooperation* (New York: Basic Books).

Bateson, P. (1983) 'Rules for Changing the Rules', in D. S. Bendall (ed.), *Evolution from Molecules to Men* (Cambridge University Press) pp. 483–507.

Bateson, P. (1988a) 'The Biological Evolution of Cooperation and Trust', in D. Gambetta (ed.), *Trust* (Oxford: Blackwell).

Bateson, P. (1988b) 'The Active Role of Behaviour in Evolution', in M. W. Ho and S. W. Fox (eds), *Evolutionary Processes and Metaphors* (Chichester: Wiley) pp. 191–207.

Boyd, R. and Richerson, P. J. (1985) *Culture and the Evolutionary Process* (Chicago: University of Chicago Press).

Chase, I. D. (1980) 'Cooperative and Non-cooperative Behaviour in Animals', *American Naturalist*, vol. 115, pp. 827–57.

Davies, N. B. (1985) 'Cooperation and Conflict among Dunnocks', *Prunella modularis*, in a Variable Mating System', *Animal Behaviour*, vol. 33, pp. 628–48.

Dawkins, R. (1986) *The Blind Watchmaker* (Harlow, Essex: Longman).

Foley, R. (1987) *Another Unique Species* (Harlow, Essex: Longman).

Hamilton, W. J. (1964) 'The Genetical Evolution of Social Behaviour. I & II', *Journal of Theoretical Biology*, vol. 7, pp. 1–52.

Hinde, R. A. (1987) *Individuals, Relationships and Culture: Links between Ethology and the Social Sciences* (Cambridge: Cambridge University Press).

Hopkins, K. (1980) 'Brother–sister Marriage in Roman Egypt', *Comparative Studies in Society and History*, vol. 22, pp. 303–54.

Houston, A. I. and Davies, N. B. (1985) 'The Evolution of Cooperation and Life History in the Dunnock, *Prunella modularis*', in R. M. Sibly and R. H. Smith (eds), *Behavioural Ecology* (Oxford: Blackwell) pp. 471–87.

Huxley, T. H. and Huxley, J. (1947) *Evolution and Ethics 1893–1943* (London: Pilot Press).

Kitcher, P. (1985) *Vaulting Ambition* (Cambridge, Mass.: MIT Press).
Kitcher, P. (1987) 'Confessions of a Curmudgeon', *Behavioral and Brain Sciences*, vol. 10, pp. 89–99.
Maynard Smith, J. (1976) 'Group Selection', *Quarterly Review of Biology*, vol. 51, pp. 277–83.
Ruse, M. (1986) *Taking Darwin Seriously: a Naturalistic Approach to Philosophy* (Oxford: Blackwell).
Taylor, M. (1976) *Anarchy and Cooperation* (New York: Wiley).
Trivers, R. (1985) *Social Evolution* (Menlo Park, Calif.: Benjamin/Cummings).
Wade, M. J. (1978) 'A Critical Review of the Models of Group Selection', *Quarterly Review of Biology*, vol. 53, pp. 101–14.
West-Eberhard, M. J. (1975) 'The Evolution of Social Behavior by Kin-selection', *Quarterly Review of Biology*, vol. 50, pp. 1–33.
Westermarck, E. (1891) *The History of Human Marriage* (London: Macmillan).
Williams, B. A. O. (1978) 'Conclusion', in G. S. Stent (ed.), *Morality as a Biological Phenomenon* (Berlin: Dahlem Konferenzen) pp. 309–20.
Williams, B. (1983) 'Evolution, Ethics, and the Representation Problem', in D. S. Bendall (ed.), *Evolution from Molecules to Men* (Cambridge: Cambridge University Press) pp. 555–66.
Wilson, D. S. (1980) *The Natural Selection of Populations and Communities* (Menlo Park, Calif.: Benjamin/Cummings).
Wilson, E. O. (1978) *On Human Nature* (Cambridge Mass.: Harvard University Press).
Wuketits, F. M. (ed.) (1984) *Concepts and Approaches in Evolutionary Epistemology: Towards an Evolutionary Theory of Knowledge* (Dordrecht: Reidel).
Zangwill, O. L. (1987) 'Hypnosis, experimental', in R. L Gregory (ed.), *The Oxford Companion to the Mind* (Oxford: Oxford University Press) pp. 328–30.

12 The First Director-General of UNESCO

W. H. G. Armytage

Huxley's own account (Huxley, 1973) of his appointment as secretary to the committee drafting a constitution for the United Nations' Educational, Scientific and Cultural Organisation reads like the précis of a 'Yes, Prime Minister' script (Lynn and Jay, 1986). For he was 'casually asked' in the spring of 1945 by the Permanent Secretary of the newly established Ministry of Education, Sir John Maud, on the steps of Lancaster House, where the preparatory commission for drafting the charter and outlining the scope of UNESCO had been taking place. For the existing secretary, Sir Alfred Zimmern, had fallen ill and a severe operation would incapacitate him for too long a time for the post to be left vacant. Huxley promised to think it over, but Maud acted very quickly indeed to twist his arm, arranging for him to dine with the newly appointed Minister that very evening and later to meet a majority of the preparatory committee (Huxley, 1973).

It was not the first such invitation he had received, for five years earlier, in August 1940, the recently retired editor of *Nature* had asked him to take over the chairmanship of the newly-founded Division of the British Association that had been established just before the war to promote the Social and International Relations of Science (Armytage, 1957). In so doing, the editor, Sir Richard Gregory, was probably influenced by Huxley's remarkable letter in *The Times* of the previous September, calling for a statement of long-term war aims on the part of the government and suggesting 'an incipient federalization of Western Europe' which would include:

the provision of large scale educational and research inter-change . . . interavailability of medical and other professional qualifications . . . a population section to deal with migration; the pooling of tropical colonial possessions, with the eventual estab-

186

lishment of an international administrative staff. (*The Times*, September 1939).

During the war Huxley addressed the American Association for the Advancement of Science which, unlike its British counterpart, continued to meet, suggesting that every country, especially in the USA, 'should begin constructing planning organisations on a large enough scale to function as a social brain and not a mere ganglion in order to ensure that any first step which we may be able to take after the war will be a step in the right direction'. (Clark, 1968).

As a former politics don, with strong Christian sympathies, at University College, Oxford, Maud must have admired Huxley's capability not only to give authoritative simplifications of difficult issues at every level (from his Romanes lecture at Oxford in 1943, ethicising his grandfather's harsh view of the evolutionary process to his successful BBC interviews with the pundits like Hyman Levy and the frankly popular – as in the Brains Trust) but also, as a Civil Servant, he must have admired the astonishing rapidity, in the days before tape-recorders and booklets, of Huxley's production of papers firming up what he had said. This latter facility – so essential in a good Civil Servant – was ironically nearly fatal in the case of UNESCO. For, as Huxley himself confessed, after he had taken up his post, he decided to take a fortnight off to clarify his ideas. In that time he dashed off a sixty-page booklet on the purpose and philosophy of UNESCO, as he envisaged it operating, far beyond the mere promotion of mutual aid, the spread of scientific ideas and cultural exchange, and help to the educational systems of backward countries, beyond and above even academic philosophies or religious doctrines. Rather, as he put it

UNESCO must work in the context of what I called *Scientific Humanism*, based on established facts of biological adaptation and advance, brought about by means of Darwinian selection, continued into the human sphere by psycho-social pressures and leading to some kind of advance, even progress, with increased human control and conservation of the environment and of natural forces. (Huxley, 1973)

This booklet was presented to the commission and ordered to be printed as an official document under the title *UNESCO: Its Purpose and its Philosophy* (Huxley, 1947)

As Huxley put the case in 1946,

Science and the scientific way of thought is as yet the one human activity which is truly universal. There is no single religious, aesthetic or political way of thought which is as yet universal. We want, therefore, to encourage this universality of scientific thought and through it help to build the basis of general universalism. (Huxley, 1946)

Hence he wished UNESCO to ensure that 'the scientific butter is spread adequately thickly over the bread of the surface of the earth' (Huxley, 1946).

But there were others who did not want to butter up the world with scientific humanism. These included the poet T. S. Eliot, The Master of Balliol (then Labour's Lord Lindsey), the chairman of the University Grants Committee, Sir Walter Moberly, the first chairman of the new Ministry of Education's Advisory Council, Sir Fred Clarke, and members of the so-called Moot. This had been meeting regularly throughout the war to ensure the perseverance of Christian principles in education. Moberly even thought that science could be taught from a Christian viewpoint. Eliot was later to sum up their feelings in *Notes towards the Definition of Culture* (1948) which he described as resting on the twin pediments of Christianity and Graeco-Roman civilisation. Other sympathisers with the Moot's quiet pressure in high places were members of the Institute of Intellectual Cooperation like Sir Ernest Barker, whom as principal at King's College, London, had once clashed with Huxley when he had been one of his professors. As one of the original supporters of the *Christian News Letter* (Kojecky, 1971) in 1939, Barker found Huxley's evolutionary humanism distasteful, despite the fact that Huxley had, in his Romanes lecture at Oxford in 1943, declared that man was now 'able to inject his ethics into the heart of evolution' (Huxley and Huxley, 1943). That this also involved contraception in planning man's destiny was almost as offensive to the Communist States, as to the Roman Catholics.

So a slip was inserted in the published copies stating that *UNESCO: Its Purpose and its Philosophy* was an expression of Huxley's personal views. Yet, despite American lobbying for Francis Biddle, the lawyer who had replaced the original nominee, Archibald Macleish, who had led the American delegation, Huxley was elected by 22 votes to 3 with two abstentions. But the Americans steadfastly refused to countenance Huxley's tenure of the post for the full six years, and secured its reduction to two.

From the close of the year 1946, when Huxley began to woo the existing international organisations, like the Council for Scientific Unions and others like the International Social Science Council, to come under UNESCO's wing, till the close of the year 1948, the pace he set was so hot that Clark called it nerve-racking. It was motivated, one might say motored, by what later became known as buzz-groups or synectics: groups where members could speak their minds once a month in the relaxing atmosphere of the Hotel Majestic's dining-room. Also monthly were the cocktail parties for everyone (Clark, 1968).

Huxley's brother Aldous had put his own experience of the old Institute for Intellectual Cooperation at his brother's disposal on 18 March 1946, when he warned him in a letter that the two major obstacles in the way of intellectuals doing anything in groups, and as groups, were (a) egotism, first personal and then national and professional, and (b) the mind's infinite capacity for irrelevance. 'People either wanted to talk about themselves, their job or their country', he wrote,

> or else they couldn't stick to the point at issue but wandered off into interstellar space. Because of these all too human failings absolutely nothing was achieved by a congress which lasted for days and must have cost the League of Nations a tidy penny. The truth is, I suspect, that you can't have a successful congress unless the subject under discussion is concrete and strictly limited and unless there is at least professional solidarity between the delegates. (Huxley, A., 1946)

At the beginning of 1946 Zimmern's staff had consisted of a dozen people in a few rooms in an empty house in Grosvenor Square, whom he hoped to expand to fifty or sixty. Then he fell ill. After Huxley took over, by the early summer of 1946 a staff of 130 were occupying two Victorian houses in Belgrave Square. Five months later in November 1946, 225 moved to the Hotel Majestic in Paris where the Peace Conference was held after the First World War. But there was no peace there after late 1946, for Zimmern had returned to conduct a whispering campaign against the man who had supplanted him labelling (or libelling) him as a Communist, to prejudice his chances of being appointed as Director.

On hearing of this, Huxley confronted the Board with the choice of either removing Zimmern or accepting his (Huxley's) resignation. So the Board found other employment for Zimmern who had the consolation of knowing that the Board had elected his supplanter for only two years instead of the constitutional five. 'Apparently', wrote Huxley in retrospect, 'some of Zimmern's mud had stuck, or perhaps some members of the Executive Board mistrusted my humanistic attitude' (Huxley, 1973). The gullible probably believed the accusation that Huxley was a Communist in view of Huxley's free association with J. D. Bernal and the appointment of Dr Joseph Needham – the man who really put the S into UNESCO – as his Director of Science (Huxley, 1973).

Though Huxley secured Zimmern's transfer elsewhere, the damage done by these accusations seems to have reverberated in the United States, for, in July 1947, the Texas branch of the American Legion was baying so insistently for Huxley's resignation, on the grounds of his atheism, that he had to reply that he considered himself a religious man, though he did not subscribe to a theistic interpretation of the religious spirit. Citing the Buddhists as an example of the non-theistic religions, he invited any Texas member of the American Legion to discuss the finer points of the question with him (Huxley, 1945).

Nevertheless, in Paris Huxley was described as infecting others with his ebullience for, as Crowther said, the effect of becoming Director was striking. His load of depression was lifted, his eyes sparkled, he strode out confidently, and his reactivated mind poured out ideas in its characteristically scintillating manner (Crowther, 1970). When by the end of 1947 the secretariat numbered more than 600, Huxley confessed to feeling the same mixture of surprise, pride and alarm over the astonishing growth of my charge, as did the giant children in H. G. Wells's *Food of the Gods* (Sewell, 1972).

But, a year later, at the general conference in Beirut, the American delegates reiterated their point of view that his term of office, having run for the stipulated two years, should cease (Clark, 1968).

But they did not quite carry the day, since the Conference implemented Joseph Needham's proposal that UNESCO should undertake a *History of Mankind*, stressing scientific and cultural progress rather than purely military and political events. And most aptly the chosen editor-in-chief was Ralph Turner who had, two decades earlier, published a biography of the first member of Parliament for Sheffield, who had a scheme for spreading science and technology around the world.

That spread had begun with the absorption of the membership of the International Conference for Scientific Unions and the establishment of new international unions in mathematics, biochemistry, crystallography, physiology, mechanics and the history and philosophy of science. By 1950, UNESCO's appropriation, in dollars, to education had doubled (from 884373 to 1624195), that to the natural sciences had grown by 70 per cent (from 498905 to 700085); the social sciences had more than quadrupled (from 82976 to 351865), that to cultural activities had increased by less than a half (from 364948 to 619197) and that to mass communication had doubled from 469032 to 969230 (Brunauer, 1945).

Huxley left office on a high note, his second and last report unrepentedly emphasising the need to balance population against resources – one of the themes in his sixty-page booklet on *UNESCO: Its Purpose and its Philosophy* which the Preparatory Commission had rejected (Huxley, 1947). This he had kept alive, at the first UNESCO Conference in Paris in 1946, by getting agreement for a study of the cultural problems caused by population increase, or decrease, that were likely to lead to war.

But, in his second and final annual report as Director-General, he skillfully reversed this causation chain by pointing out that even a war was a less inevitable threat to civilisation than population increase. Equally skillfully, he referred his proposal to educate people about the need to have a balance between resources and population through the convention of a conference of individuals unfettered by governmental or national ties. This conference he wished to be sponsored by four bodies other than, but with, UNESCO (WHO, FAO, ECOSOL and ILO). The inevitable delaying tactic – postponement – was advanced most reasonably in the form of the suggestion that UNESCO should wait until the results of the 1950 Census had come in. This prevented such a conference being held until 1954. But it was then a bigger and more ambitious venture – nothing less than the first of what were to be regular World Population Conferences. There had been one 27 years earlier, at which, needless to say, Huxley had spoken, but it was really a private affair (Huxley, 1949).

Now, however, the 1954 Conference was held in Rome and delegates from Catholic countries were among those from 58 governments who attended. All the delegates, however, were received by the Pope, whose address was printed in an English translation in the *Eugenics Review* in April 1955.

By the time the *Eugenics Review* had printed this, Huxley had been a member of the British Commission on UNESCO for seven years (effected immediately after his retirement as Director-General at the end of 1948). In this new capacity he helped to decide what projects the British would support. Moreover, as Vice-President of the Commission dealing with UNESCO's cultural *History of Mankind* he obtained Jacquetta Hawkes and Leonard Wooley as co-authors of the first volume.

So, when he was hailed as 20 years ahead of his time for emphasising the relation between population and the destruction of the environment (Symonds and Carder, 1973), we should not ignore the enormous amount of travel, investigation, conversation and reporting that he did in the 1950s and 1960s, that climaxed in 1970 with one Conference on a long-term programme on *Man and the Biosphere*, and, with another on *World Wildlife* (of which he initiated the formation), when the Duke of Edinburgh stressed the need for everyone to participate in the preservation of the environment, and Huxley was preached.

In the service of the agency founded and affiliated to UNESCO when Huxley was Director-General – the International Union for the Conservation of Nature (IUCN) – and its associated body, the World Wildlife Fund, he travelled to Manila, Sarawak, Borneo, Java, Bali, Singapore, Thailand and the emergent African States.

Huxley's prescript is still relevant, perhaps more so than ever before. For in thirteen years' time (when the century ends) the world's population will have doubled from what it is now (1.7 to 3.3 billion). More importantly, half of this increase will be in Asia, a quarter in Africa and a tenth in Latin America (Food 2000, 1987). And in 1987, with the US estranged from UNESCO and the United Kingdom only in a watchdog role, a World Commission on Environment and Development has argued for what it calls 'resource literacy through technical skills' – the E and S of UNESCO. So, let us hope that in 1987 when the general Conference will have to elect a successor to the present Director-General, Mr M'Bow, the Singalese who was elected in 1974 before Huxley's death – though not, of course, Huxley's successor – will be mindful of this, and of the need to win back both the United States of America and the United Kingdom to full membership once more, for the present situation is intolerable.

References

Armytage, W. H. G. (1957) *Sir Richard Gregory* (London: Macmillan) p. 177.
Brunauer, E. C. (1945) 'International Council of Scientific Unions', *U.S. Department of State Bulletin*, vol. 13, pp. 371–6.
Clark, R. W. (1968) *The Huxleys* (London: Heinemann) pp. 276–7, 316–7, 322.

Crowther, J. G. (1970) *Fifty Years with Science* (London: Barrie and Jenkins) p. 89.

Eliot, T. S. (1948) *Notes Towards the Definition of Culture* (London: Faber & Faber).

Food 2000: Global Policies for Sustainable Agriculture (1987) *Report to the World Commission on Environment and Development* (London: Z Books).

Huxley, A. (1946) Letter to J. S. Huxley, in Grover Smith (ed.), *Letters of Aldous Huxley* (London: Chatto & Windus 1969) pp. 538–9.

Huxley, J. S. (1945) 'Science and the United Nations', *Nature*, vol. 156, pp. 553–6.

Huxley, J. S. (1946) 'The Future of UNESCO', *Discovery*, vol. 7, pp. 72–3.

Huxley, J. S. (1947) *UNESCO: Its Purpose and its Philosophy*, UNESCO C/6 15 September 1946 (Washington: Public Affairs Press).

Huxley, J. S. (1949) 'UNESCO: A Year of Progress', *United Nations Bulletin (New York)* vol. 6, pp. 28–30.

Huxley (1973) *Memories II* (London: George Allen and Unwin).

Huxley, J. S. and Huxley, T. H. (1943) *Evolutionary Ethics* (Oxford: Oxford University Press).

Kojecky, R. (1971) *T. S. Eliot's Social Criticism* (London: Faber and Faber) p. 174.

Lynn, J. and Jay, A. (1986) *Yes, Prime Minister: The Diaries of the Rt. Hon. James Hacker* (London: BBC Publications).

Sewell, J. P. (1972) *UNESCO and World Politics: Engaging in International Relations.* (Princeton University Press) p. 91.

Symonds, R. and Carder, M. (1973) *The United Nations and the Population Question 1945–1970* (London: Chatto and Windus) p. xii.

Note: Those seeking information on the wider context of the origins of UNESCO should consult:

C. S. Ascher (1951) *Program-making in UNESCO 1946–1951* (Chicago: Public Administration Clearing House).

L. G. Camery (1949) *American Background of the United Nation's Educational, Scientific and Cultural Organisation*. PhD, Stanford.

F. R. Cowell (1966) 'Planning the Organisation of UNESCO, 1942–1946: a Personal Record', *Journal of World History*, vol. 10, pp. 210–36.

Y. T. Feng (1953) *Analysis of the Impact of Several Different Concepts of International Co-operation upon the Establishment and Development of the United Nations Educational, Scientific and Cultural Organization during its First Six Years*. PhD, Denver.

J. Havet (1948) 'Is There a Philosophy of UNESCO?', in L. Bryson, L. Finkelstein and R. MacIver (eds), *Symposium on Science, Philosophy and Religion 1947* (New York: Harper).

B. Karp (1951) *Development of the Philosophy of UNESCO*. PhD, Chicago.

W. G. Leyland (1946) 'The Background and Antecedents of UNESCO', *Proceedings of American Philosophical Society*, vol. 90, pp. 256–9.

R. McKeon (1948) 'A Philosophy for UNESCO', in *Philosophy and Phenomenological Research*, vol. 3, June 1948.

13 Julian Huxley and Eugenics

David Hubback

Throughout his career Huxley was a great believer in Eugenics. Of Dean Inge's dictum that,

> Eugenics is capable of becoming the most sacred ideal of the human race, as a race: one of the supreme religious duties,

Huxley said in his 1936 Galton lecture

> I entirely agree with him. Once the full implications of evolutionary biology are grasped, eugenics will inevitably become part of the religion of the future, or of whatever complex of sentiments may in the future take the place of organized religion. (Huxley, 1936)

He ended his second Galton lecture in 1962:

> If, as I firmly believe, man's role is to do the best he can to manage the evolutionary process on this planet and to guide its future course in a desirable direction, fuller realization of genetic possibilities becomes a major motivation for man's efforts, and eugenics is revealed as one of the basic human sciences. (Huxley, 1962)

These seem extraordinary remarks to modern ears after the plunge in the public esteem of eugenics following the Nazis' perversion of eugenics to justify their creed of a master race, and the growing post-war hatred of racism and Big Brother authoritarian government. With 'Eugenics' becoming a dirty word, the American Eugenics Society changed its name, in 1972, to the Society for the Study of Social Biology and our own Eugenics Society kept its name only after long debate, and ceased to be a pressure group. As a result the admirable broad-ranging *Eugenics Review* was wound up in 1968 and the Society keeps a comparatively low profile except for these annual symposia. Even now, when we would have expected passions to have cooled, there are diatribes against eugenics in general and against past eugenists, including Huxley, in particular. One of the most aggressive is that of Germaine Greer in her book *Sex and Destiny* (1984) where she writes

> Practical eugenics denies all the values which justify our civilization. When we have a clearer idea of our own ignorance we shall see that eugenics is more barbarous than cannibalism and far more destructive.

There is of course no need for me to justify to this audience continuing work on eugenics, given the huge strides that have been made in genetics and genetic engineering since the 1950s and the bright prospects for further early major advances. But Huxley's enthusiasm for eugenics throughout most of his life was remarkable. Although it is a little odd that eugenics hardly figures in his autobiography, *Memories* (Huxley, 1970), he gave a lot of time to the Eugenics Society, he wrote extensively about the subject and was the only man to have given two Galton lectures. Where did his enthusiasm come from and how far was it justified?

HUXLEY'S ENTHUSIASM FOR EUGENICS

I am not qualified to answer the second question although I shall express some personal views. At this stage all I need say is that some of his scientific contemporaries thought he built too much on the comparatively little that was then known about human genetics; others that he at least ensured that the right questions were discussed. I will, however, try from an historical point of view to answer the first question.

Huxley's enthusiasm was that of a polymath scientist who fell in love with general ideas – especially in politics and sociology. He was very much the grandson of T. H. Huxley. He liked taking the broad synoptic view. He was an excellent simplifier of complicated scientific theories, as is shown by the work he did with H. G. and G. P. Wells on the popular encyclopaedic *Science of Life*, published in 1929. He was much influenced, when a young professor in 1913 at the Rice Institute in Texas, by his assistant H. J. Muller who was working on the effects of radiation on germ plasm. At an early stage in his career he came to realise the importance of genetics as a discipline basic to an understanding of evolution. After the First World War his work on zoology at Oxford and King's College, London, greatly extended his scientific ideas and widened his circle of friends, mostly 'progressive' in their views. He became interested in population problems and birth control. With his bounding energy he had become by the mid-1920s an active member of the Eugenics Society, which provided an excellent forum for discussion between natural scientists, social scientists and social reformers.

It was, of course, by chance that it was C. P. Blacker, the future General Secretary of the Eugenics Society who picked Huxley up off the floor after he collapsed during a Zoology class he was giving in Oxford in the summer of 1919 – a collapse which started the first of the series of depressions from which he suffered throughout his life, and which his widow Juliette Huxley (1986) has so sympathetically described in her book *Leaves of the Tulip Tree*. But Blacker, who shared Huxley's passion for bird-watching, admired him and worked with him fruitfully at the Eugenics Society and elsewhere.

THE EUGENICS SOCIETY

A brief description of the Eugenics Society of the 1920s and 1930s may be helpful here. It was a typically English institution, bringing together the eminent, and not so eminent, from many professions to discuss the implications of human genetics for social policy. Demography, the economic consequences of a falling birth rate, the need for family allowances and for genetic counselling, safer and more effective birth control, artificial insemination by donor (AID) and sterilisation were all discussed. Peter Medawar later called Eugenics 'the political arm of human genetics' (Medawar, 1977).

The Society certainly debated political aims and at one time had a propaganda secretary. It pressed for voluntary sterilisation to become legal. Scientific research was encouraged but it was not the main task of the Society. After 1930, thanks to the bequest of a wealthy Australian sheep farmer (who thought he had been born of unsound parents, and hence declined to marry) the Society was well funded, but it did not have enough money for extensive research.

Sidney Webb and Maynard Keynes were early members – Keynes being Treasurer of the Cambridge branch in 1914 and serving on the Council up to his death in 1946. The Society attracted Fabians as well as leading scientists (J. B. S. Haldane and Lancelot Hogben were active, but outside the Society partly because of their dislike of Blacker). Josiah Stamp, William Beveridge and Richard Titmus joined, as did demographers such as Carr-Saunders, David Glass and Kuczynski. Medical doctors included Lord Dawson of Penn and Lord Horder., The case for Family Allowances was pursued by R. A. Fisher from a strictly eugenic point of view, and by Eleanor Rathbone, Mary Stocks and Eva Hubback, as social reformers.

Huxley became active in the Society when Leonard Darwin, the fourth son of Charles Darwin, was still president, a post he occupied from 1911 to 1928. Huxley himself was soon on the Council of the Society and became a vice-president.

All in all, the progressive establishment was very much at home in the Society whose influence was increased by its journal, *The Eugenics Review*, noted both for its content and its handsome typography.

Eugenics was indeed an ideal subject for Julian Huxley. It played an important part in his faith in scientific humanism and in his inborn optimism about the possibility of a better life for mankind. Eugenics seemed to offer a practical way of helping Man to take charge of his destiny by controlling his own evolution.

HUXLEY'S CHANGING POLITICAL VIEWS

Huxley in the 1920s was a middle-of-the-road liberal with progressive views on birth control and the dangers of excessive world population growth. He was, however, still sufficiently steeped in traditional eugenics to advocate in 1931 that the granting of unemployment pay should be conditional on the recipient having no more children.

Infringement of this order could probably be met by a short period of segregation in a labour camp. After three to six months' separation from his wife he would be likely to be more careful the next time. (Huxley, 1931)

Thereafter, with continuing high unemployment at home and the rise of Fascism abroad, Huxley moved to the left, as did most of his scientific contemporaries interested in genetics. J. B. S. Haldane and Hyman Levy (with whom Huxley had BBC discussions which helped to establish his reputation as an outstanding broadcaster) became communists. Hogben, then Professor of Social Biology at the London School of Economics, moved left too. Huxley, after a visit to Russia in 1932, where he was convinced that scientists and scientific methods were properly esteemed, became interested in State planning, and from 1932 was the resident scientist at Political and Economic Planning (PEP) – a non-political body concerned with planning problems. Here he met Max Nicholson, then a young economist and another outstanding ornithologist, who worked with Huxley later on world population problems. Huxley's association with PEP greatly influenced him when he wrote *If I were Dictator* in 1934 – a long essay mainly about the case for a planned society based on scientific humanism. He wanted a scientific society ruled by an enlightened elite.

Huxley then shared the left-wing political consensus which predominated in the 1930s among his generation of scientists and intellectuals. While not active politically, except as an essayist, he became highly critical of capitalism and nationalism, considering them both dysgenic. His thinking on eugenics led him to stress the need to improve the social environment, particularly the diet of the working class, which had been shown in 1936, by Sir John Boyd Orr (1936) to be gravely inadequate.

It is not surprising that Huxley became highly critical of the crude, so-called 'mainline' eugenic beliefs which had grown up in the USA, and to a lesser extent in Britain, in the first part of the twentieth century. Mainline eugenists opposed birth control on the grounds that it would accentuate the dysgenic differential fertility of the feckless poor who would not use it, as opposed to the more careful upper classes who would. Leonard Darwin thought birth control would be 'racially devastating'.

Early eugenists pressed for the sterilisation of mentally defective people suffering from inherited diseases and of members of 'problem families'. There was no compulsory sterilisation in Britain but 24 States in the USA had taken such powers by 1929 and over 6000 sterilisations had been carried out in California alone. Immigration laws too were designed to prevent the protestant North European inhabitants of the United States being swamped by Catholic immigrants from Mediterranean countries – or by the yellow and Negro races.

The political pressures were clearly much greater in the United States than in Britain, where there had been no attempt to introduce eugenic policies such as sterilisation. But it was British scientists, such as Huxley, R. A. Fisher, J. B. S. Haldane and Lancelot Hogben, who, for a variety of motives, argued that the old eugenics were based on bad science. The concept of race made no biological sense, and while sterilisation could reduce some dominant hereditary diseases, such as Huntingdon's chorea, it could have little effect on the far more numerous recessive-gene diseases where genetic counselling would be more helpful. It seemed to Huxley and Haldane that sterilisation was being used against the poor, who were falsely equated with the feeble-minded.

GENETIC DECLINE IN BRITAIN?

Huxley, like many of his contemporaries, had two major concerns. The first (in spite of the lack of figures after the 1911 census to show what was really happening to fertility) was that the differential fertility of the feckless poor and the responsible prosperous classes, especially the professional classes, was leading to a genetic decline in Britain. Second, because of social welfare, this decline was being accentuated by natural selection not being allowed to weed out defective stock resulting from genetic mutations which were usually harmful, nor was natural selection being replaced by artificial selection, as with the breeders of cattle and plants. He concluded that there was a real danger that humanity would gradually destroy itself from within.

The problem for Huxley was how could the eugenic policies required be reconciled with the liberal socialism he advocated, which was designed to be egalitarian and to protect the weak in society?

HUXLEY'S FIRST GALTON LECTURE

In his first Galton lecture of 1936 Huxley concluded that the best hope lay in measures to improve the social environment by levelling up the standard of living of the poor, together with eugenic policies designed to discourage bad human stock reproducing while encouraging good stock to multiply.

In passing it is worth noting that the possible dysgenic differential fertility of social classes and of races has always been of concern to eugenists, although the fears about particular societies at particular times have often been shown not to be justified. In the Britain of the 1930s, the birth rate of all classes was falling and the main worry became the prospect of a declining population. There is still concern about differential fertility today, particularly of immigrant groups, but the most recent figures show these fears have been exaggerated. Most immigrant groups after a time tend to follow the customs of their adopted country, and there is now not much difference between the fertility of the different social classes in Britain. Social groups IV and V are admittedly about 16 per cent more fertile than the other groups, which are bunched together, but these two lowest groups are rapidly declining as a proportion of the total population (*Population Trends* No. 41, 1985).

EUGENICS – A SOCIAL SCIENCE

Huxley was a natural scientist who wanted to be a social scientist as well. He argued in his Galton lecture that Eugenics was one of the social sciences, rather than a natural science, and a young one at that. But it did not follow that Eugenics was not truly scientific; only that it was more difficult and complex, being determined by a multiplicity of partial causes rather than by one simple cause. Huxley admitted that the social scientist is part of his own material and therefore cannot escape bias. (Eugenists have indeed often been accused of self-delusion, particularly when defining the desirable qualities in mankind to be encouraged in terms of the values of their time and class; typically those of upper-middle-class intellectuals with an Oxbridge background.)

Huxley saw Eugenics as a particular aspect of the study of man in society. Until the social environment had been levelled up for all, it would be difficult to evolve effective eugenic policies. Huxley maintained that biology shows 'the inherent diversity and inequality of man'. People in the same socio-economic class scored across a wide range in IQ tests. This could be accounted for only by a variation in natural ability. Huxley predicted that even if environmental disparities were eliminated

> the genetically flawed core of the 'social problem' group would remain and the professional classes would be revealed as a reservoir of superior germ plasm of high average level notably in regard to intelligence.

Huxley, however, recognised that there was a great spread of ability in every class and that a sizeable number of people in the lowest group were at least as intelligent as a sizeable number of people in the highest group.

As a first step it was necessary to ensure adequate diet, health care, housing and education for all. But widespread social improvements were not enough. Huxley insisted that a social system based on private capitalism and the nation state was bound to be dysgenic since it failed to make use of the genetic abilities of the poor and was likely to lead to war, the ultimate dysgenic.

NEGATIVE AND POSITIVE EUGENICS

Huxley was convinced that no really rapid progress would come from encouraging the reproduction of one class or one race against another. Striking and rapid eugenic results could only come by the virtual elimination of the few lowest and truly degenerate types, and a high multiplication rate of the few highest and truly gifted types. The first objective could in time be achieved by contraception and, where necessary, by abortion and sterilisation, the second by artificial insemination (AID) using the sperm of eminent highly gifted men. AID had made it possible to separate sexual love and reproduction.

Huxley regarded the capitalist system as encouraging anti-social competitive characteristics which were dysgenic. The aim should therefore be to move towards a more cooperative pattern of economic and communal life. He believed that:

> there is no doubt that genetic differences of temperament, including tendencies to social or anti-social action, cooperation or individualism, do exist, nor that they could be bred for in man as man has bred for tameness and other temperamental traits in domestic animals, and it is extremely important to do so. If we do not, society will be continuously in danger from the antisocial tendencies of its members.

This last statement seems highly questionable but his faith was shared by a group of 22 leading British and American biologists including Haldane, Hogben, Needham and Waddington who, with Huxley, just as war was breaking out in 1939, issued a manifesto on 'Biology and Population Improvement'. The manifesto stated:

> The most important genetic objectives, from a social point of view, are the improvement of those genetic characteristics which make (a) for health, (b) for the complex called intelligence, and (c) for those temperamental qualities which favour fellow-feeling and

social behaviour rather than those (today most esteemed by many) which make for personal 'success' as success is usually understood at present.

A more widespread understanding of biological principles will bring with it the realization that much more than the prevention of genetic deterioration is to be sought for, and that the raising of the level of the average of the population, nearly to that of the highest now existing in isolated individuals, in regard to physical well-being, intelligence and temperamental qualities, is an achievement that would – so far as purely genetic considerations are concerned – be physically possible within a comparatively small number of generations. Thus everyone might look upon 'genius', combined of course with stability, as his birthright. (*Nature*, 1939)

This manifesto is said to have been drafted mainly by Muller, but Huxley must have had a lot to do with it as it closely follows his Galton lecture. Its acceptance by so many biologists, some of whom were highly critical of Huxley as not really understanding genetics, may have been an expression of faith to offset their gloom at the outbreak of war. But it can be argued that if Huxley was naïve in signing it, so were his critics.

EUGENICS 1939–1962

Huxley's faith in the value of Eugenics, as he had defined it, remained constant in the quarter-century up to his second Galton lecture in 1962, even though his political views moved, with those of most of his contemporaries, towards the centre.

The spate of his well-written books and articles, often with some bearing on evolution and eugenics, continued. He was active in emphasising the dangers of excessive world population growth and in putting up barriers against the rapid degradation of the world environment through helping to start the International Union for the Conservation of Nature in 1948 and the World Wild Life Fund in 1961.

His Galton lecture of 1962, although on similar lines to that of 1936, had to take account of many changes. Eugenics, because of the Nazis, had become a dirty word. The membership of the Eugenics Society fell sharply and it ceased to be a preeminent forum as in the 1930s. Human genetics had, however, made rapid advances, culminating in the discovery of DNA. Contraceptive methods had improved and their use was much more widespread, although the pill had still to become popular and vasectomy was not introduced in Britain until the mid-1960s. Then it was Blacker, with the help of the Simon Population Trust, who took the lead. AID, using deep-frozen sperm banks, was just starting up. Eugenic counselling, to help husband and wife decide about possible inherited risks the children they might conceive would run, was developing.

The nuclear bomb had raised great anxiety about the long-run dysgenic effects of radiation, although the detailed research carried out at Hiroshima did not reveal great immediate damage. The other bomb, of rapidly increasing world population, became of increasing concern to Huxley. In 1946 Huxley hoped that UNESCO, of which he was the first director-general, might encourage the provision of birth-control centres world-wide.

Huxley himself was concerned that there was genetic decline in progress because of the increasing number of genetic defectives being kept alive and because of the new crop of harmful genetic mutations to be expected from radio-active fall-out. But he remained optimistic that the right social and eugenic policies could change the negative signs to positive.

THE SECOND GALTON LECTURE

In his 1962 Galton lecture Huxley argued it would be possible to lighten man's growing genetic load by discouraging genetic defectives from breeding, by reducing man-made radiation to a minimum, and by slowing down human over-multiplication in general and differential fertility of various classes and nations in particular. It would then be possible to develop positive eugenics by introducing graded family allowances to help specially able parents to have more children, and by much greater use of AID, or Eugenic Insemination as Huxley called it, by which parents could select the qualities they wished their children to have by drawing on a deep-frozen sperm bank donated by eminent individuals. Huxley argued that couples who adopted this method of vicarious parenthood would be rewarded by children outstanding in qualities admired by the couples themselves. Huxley hoped parents would opt for intelligence – initial projects for sperm banks called for Nobel prize-winners as donors. On the supposition that genetic intelligence is polygenically determined, and that its distribution follows a normal symmetrical curve, the raising of the mean genetic IQ of a population by 1 1/2 per cent would result in a 50 per cent increase in the number of individuals with an IQ of 160 or over. That, Huxley thought, would lead to a much more enlightened and efficient society.

Huxley brushed aside objections from Peter Medawar (1960, 1974 and 1977) and others that Eugenic Insemination, or other methods of improving the genetic make-up of mankind, were neither genetically nor politically feasible. The objectors argued that we simply did not know enough about human genetics. For instance, it is quite unknown to what extent non-physical human characteristics have a genetic basis. In any case, stock-breeding techniques would not work. Moreover, it would be authoritarian and misleading to persuade parents to use 'approved' sperm. Huxley retorted that it was possible to make scientific advances without knowing the precise mechanics. Darwin had worked out his theory of evolution without knowing anything about the actual mechanism of biological variation and inheritance. There was no question of stock-breeding methods. Eugenic insemination, once generally acceptable, as birth-control had become acceptable, would became an entirely voluntary parental choice from a whole range of preferred types, so that the fruitful diversity of the human species would be drawn upon. Huxley did, however, admit that he felt embarrassed when H. J. Muller suggested he should contribute to a sperm bank being set up in the USA.

Huxley finished his lecture on a rosy cloud of optimism about the future of eugenics. In the perspective of evolution, he asserted:

> . . . eugenics – the progressive genetic improvement of the human species – inevitably takes its place among the major aims of evolving man.

EUGENICS SINCE 1962

It is not for me to assess the scientific basis for Huxley's belief in eugenics. As far as I am aware, present-day geneticists, in spite of all their recent discoveries, tend to be much more cautious in their predictions than was Huxley. Public debate now is more about the ethics that should govern the use of the new discoveries, rather than about the social reform, so much debated in the 1930s, that should go with them. Today it is the moral issues raised by the Warnock report that get discussed, rather than the brave new world that might be opened up by these discoveries. Few people now share Huxley's belief in Utopia.

I can only ask how some of Huxley's own ideas have stood the test of time. AID is widely used and there are far more women heads of single-parent families wanting babies but preferring to dispense with husbands. But Eugenic Insemination itself does not seem to have made the progress Huxley hoped for. A sperm bank was set up in the late 1970s in the USA using Nobel Laureates only – with a later relaxation to scientists (but not the humanities) generally. By 1984 fifteen offspring had been produced but of course it is much too early to judge the outcome of these, and of other experiments there may have been (Kevles, 1985, p. 263).

As for negative eugenics, Huxley would have been encouraged by the prospect (thanks to recent advances towards finding out the determinative effect of human genes and combinations of genes, and to the refinement of genetic probes and amniocentesis) of significantly reducing the genetic disorders that have been estimated to occur in half to one per cent of all live births in the United States and Britain. But even if, as Professor Ruddle of Yale has predicted, we achieve the major outline of the human gene map by the year 2000 there will still be a long way to go (Kevles, 1985, p.297).

The human genetic make-up has been shown to be highly complex. There is no real division between 'healthy normal' people and the genetically diseased. It appears we are all carriers of some disease, but are very unlikely to pass on that disease (Bird Stewart, 1987). It can even be argued that recent advances in genetics have suggested that negative eugenics would probably be positively harmful in that the variability of the gene pool, on which evolutionary forces work, would be reduced. But I doubt if Huxley would have agreed that it is best to leave well alone. He would have been more likely to remain optimistic about the eventual practical applications of the present great volume of genetic research.

As for world population growth, Huxley would still be worried. The world has just acquired its five-billionth inhabitant, and while the recent steep rate of increase is slightly flattening out, we shall have six billion by the year 2000 and will be lucky to stay much below an eventual maximum of ten billion (double today's figure) in the second half of the twenty-first century. There is everything to fight for in keeping that peak figure down if the quality of most human life is not to deteriorate further, thus offsetting any gain from the prospective reduction of hereditary and other diseases.

Huxley might also be worried about the differential fertility of the Europeans and North Americans as compared with that of the Latin Americans, the Africans and the oriental peoples. It seems the Europeans will only provide about 6 per cent of the world's population in fifty years' time. Indeed, they contribute less than 10 per cent now. North America looks like becoming more hispanic in race, with growing proportions too of orientals and perhaps blacks. Whether or not the resulting mix will be eugenic can be debated, but it certainly seems 500 years of European domination of the world are coming to an end.

Huxley might have been sad that the Eugenics Society is no longer a pressure group that could further his policies, but I suspect he would have very much liked to take part in this Symposium.

References

Greer, G. (1984) *Sex and Destiny* (London: Secker & Warburg) p. 293

Huxley, J. S. (1931) *What Dare I Think?* (London: Chatto & Windus).

Huxley, J. S. (1934) *If I Were a Dictator* (London: Methuen).

Huxley, J. S. (1936) Galton Lecture: *Eugenics and Society*, published 1941 in *The Uniquesness of Man* (London: Chatto & Windus).

Huxley, J. S. (1962) Galton Lecture: *Eugenics in Evolutionary Perspective*, published 1964 in *Essays of a Humanist* (London: Chatto & Windus).

Huxley, J. S. (1970) and (1973) *Memories*, vols I and II (London: George Allen & Unwin).

Huxley, Juliette (1986) *Leaves of the Tulip Tree* (London: John Murray).

Kevles, D. J. (1985) *In the Name of Eugenics* (London: Penguin Books).

Medawar, Sir P. (1960) *The Future of Man*, The Reith Lectures 1959 (London: Methuen).

Medawar, Sir P. (1974) 'The Genetic Improvement of Man', in *The Hope of Progress* (London: Wildwood House).

Medawar, Sir P. with J. S. Medawar (1977) *The Life Science*, Ch. 7 (London: Wildwood House).

Nature (1939) '*Biology and Population Improvement*. Biologists Manifesto', *Nature*, vol. 144, 16 September 1939.

Orr, Sir John Boyd (1936) *Food, Health and Income* (London: Macmillan).

Population Trends (1985) 'Differential Fertility by Social Class', *Population Trends*, No.41, August 1985, HMSO.

Stewart, J. Bird (1987) 'Genetics in Relation to Biology', *School Science Review*, June 1987, pp. 645–53.

14 The Galton Lecture for 1962: Eugenics in Evolutionary Perspective*

Sir Julian Huxley, FRS

I am honoured at having been twice asked to give the Eugenics Society's Galton Lecture. The first occasion was a quarter of a century ago, when Lord Horder was our President, and I am proud of the remarks which he and my brother Aldous made about these.

Let me begin by broadly outlining how eugenics looks in our new evolutionary perspective. Man, like all other existing organisms, is as old as life. His evolution has taken close on three billion years. During that immense period he – the line of living substance leading to *Homo sapiens* – has passed through a series of increasingly high levels of organisation. His organisation has been progressively improved, to use Darwin's phrase, from some submicroscopic gene-like state, through a unicellular to a two-layered and a metazoan stage, to a three-layered type with many organ-systems, including a central nervous system and simple brain, on to a chordate with notochord and gill-slits, to a jawless and limbless vertebrate, to a fish, then to an amphibian, a reptile, an unspecialised insectivorous mammal, a lemuroid, a monkey with much improved vision, heightened exploratory urge and manipulative ability, an ape-like creature, and finally through a protohominid australopith to a fully human creature, big-brained and capable of true speech.

This astonishing process of continuous advance and biological improvement has been brought about through the operation of natural selection – the differential reproduction of biologically beneficial combinations of mutant genes, leading to the persistence, improvement and multiplication of some strains, species and patterns

---

* The Galton Lecture, delivered in London on 6 June 1962. From *The Eugenics Review*, October 1962, vol.54, pp. 123–41. Reprinted by permission of the Eugenics Society and Mr Anthony Huxley.

of organisation and the reduction and extinction of others, notably to a succession of so-called dominant types, each achieving a highly successful new level of organisation and causing the reduction of previous dominant types inhabiting the same environment. During its period of dominance, which may last up to a hundred million years or so, the new type itself becomes markedly improved, whether by specialization of single subtypes like the horses or elephants, or by an improvement in general organisation, as happened with the mammalian type in general at the end of the Oligocene. Eventually no further improvement is possible, and further advance can only occur through the breakthrough of one line to a radically new type of organisation, as from reptile to mammal.

In biologically recent times, one primate line broke through from the mammalian to the human type of organisation. With this, the evolutionary process passed a critical point, and entered on a new state or phase, the psychosocial phase, differing radically from the biological in its mechanism, its tempo, and its results. As a result, man has become the latest dominant type in the evolutionary process, has multiplied enormously, has achieved miracles of cultural evolution, has reduced or extinguished many other species, and has radically affected the ecology and indeed the whole evolutionary process of our planet. Yet he is a highly imperfect creature. He carries a heavy burden of genetic defects and imperfections. As a psychosocial organism, he has not undergone much improvement. Indeed, man is still very much an unfinished type, who clearly has actualised only a small fraction of his human potentialities. In addition, his genetic deterioration is being rendered probable by his social set-up, and definitely being promoted by atomic fallout. Furthermore, his economic, technical and cultural progress is threatened by the high rate of increase of world population.

The obverse of man's actual and potential further defectiveness is the vast extent of his possible future improvement. To effect this, he must first of all check the processes making for genetic deterioration. This means reducing man-made radiation to a minimum, discouraging genetically defective or inferior types from breeding, reducing human over-multiplication in general and the high differential fertility of various regions, nations and classes in particular. Then he can proceed to the much more important task of positive improvement. In the not too distant future the fuller realisation of possibilities will inevitably come to provide the main motive for man's overall efforts; and a Science of Evolutionary Possibilities, which today is merely adumbrated, will provide a firm basis for these efforts. Eugenics can make an important contribution to man's further evolution: but it can only do so if it considers itself as one branch of that new nascent science, and fearlessly explores all the possibilities that are open to it.

Man, let me repeat, is not a biological but a psychosocial organism. As such, he possesses a new mechanism of transmission and transformation based on the cumulative handing on of experience, ideas and attitudes. To obtain eugenic improvement, we shall need not only an understanding of what kind of selection operates in the psychosocial process, not only new scientific knowledge and new techniques in the field of human genetics and reproduction but new ideas and attitudes about reproduction and parenthood in particular and human destiny in general. One of those new ideas will be the moral imperative of Eugenics.

*   *   *

In the twenty-five years since my previous lecture, many events have occurred, and many discoveries have been made with a bearing on eugenics. Events such as the explosion of atomic and nuclear bombs, the equally sinister 'population explosion,' the *reductio ad horrendum* of racism by Nazi Germany, and the introduction of artificial insemination for animals and human beings, sometimes with the use of deep-frozen sperm; scientific discoveries such as that of DNA as the essential basis for heredity and evolution, of subgenic organisation, of the widespread existence of balanced polymorphic genetic systems, and of the intensity and efficacy of selection in nature; the realisation that the entities which evolve are populations of phenotypes, with consequent emphasis on population genetics on the one hand, and on the interaction between genotype and environment on the other; and finally the recognition that adaptation and biological improvement are universal phenomena in life.

I do not propose to discuss these changes and discoveries now, but shall plunge directly into my subject – Eugenics in Evolutionary Perspective. I chose this title because I am sure that a proper understanding of the evolutionary process and of man's place and role in it is necessary for any adequate or satisfying view of human destiny; and eugenics must obviously play an important part in enabling man to fulfil that destiny.

As I have set forth at greater length elsewhere, in the hundred years since the publication of the *Origin of Species* there has been a 'knowledge explosion' unparalleled in all previous history. It has led to an accelerated expansion of ideas, not only in the natural sciences but also in the humanistic fields of history, archaeology, and social and cultural development, and its effects on our thinking have been especially violent, not to say revolutionary, during the quarter of a century since my previous Galton Lecture. It has led to a new picture of man's relations with his own nature and with the rest of the universe, and indeed to a new and unified vision of reality, both fuller and truer than any of the insights of the past. In the light of this new vision the whole of reality is seen as a single process of evolution. For evolution can properly be defined as a natural process in time, self-varying and self-transforming and generating increasing complexity and variety during its transformations; and this is precisely what has been going on for all time in all the universe. It operates everywhere and in all periods, but is divisible into a series of three sectors or successive phases, the inorganic or cosmic, the organic or biological, and the human or psychosocial, each based on and growing out of its predecessor. Each phase operates by a different main mechanism, has a different scale and a different tempo of change, and produces a different type of results.

Between the separate phases, the evolutionary process has to cross a critical threshold, passing from an old to a new state, as when water passes from the solid to the liquid state at the critical temperature-threshold of 0 °C and from liquid to gaseous at that of 100 °C.

The critical threshold between the inorganic and the biological phase was crossed when matter and the organisms built from it became self-reproducing, that between the biological and the psychosocial when mind and the organisations generated by it became self-reproducing in their turn.

The cosmic phase operates by random interaction, primarily physical but to a small degree chemical. Its quantitative scale is unbelievably vast both in space and time. Its visible dimensions exceed 1000 million light-years (10^{22} km), its distances are measured by units of thousands of light-years (nearly 10^{16} km), the numbers of its visible galaxies exceed 100 million (10^8) and those of its stars run into thousands of millions of millions (10^{15}). It has operated in its present form for at least 6000 million years, possibly much longer. Its tempo of major change is unbelievably slow, to be measured by 1000-million-year periods. According to the physicists, its overall trend, in accord with the Second Law of Thermodynamics, is entropic, tending towards a decrease in organization and to ultimate frozen immobility; and its products reach only a very low level of organisation – photons,

'Extraordinary. After all that publicity about artificial insemination by admired donors, not a single letter.' (Reproduced by permission of *Punch* in *The Eugenics Review*, 1962, vol. 54, p. 175)

subatomic particles, atoms, and simple inorganic compounds at one end of its size-scale, nebulae, stars and occasional planetary systems at the other.

The biological phase operates primarily by the teleonomic or ordering process of natural selection, which canalises random variation into non-random directions of change. Its tempo of major change is somewhat less slow, measured by 100-million- instead of 1000-million-year units of time. Its overall trend, kept going of course by solar energy, is anti-entropic, towards an increase in the amount and quality of adaptive organisation, and marked by the growing importance of awareness as mediated by brains. And its results are organisms – organisms of an astonishing efficiency, complexity, and variety, almost inconceivably so until one recalls R. A. Fisher's profound paradox, that natural selection plus time is a mechanism for generating an exceedingly high degree of improbability.

In the course of biological evolution, three sub-process are at work. The first (cladogenesis, or branching evolution) leads to divergence and greater variety; the second (anagenesis, or upward evolution) leads to improvement of all sorts, from detailed adaptations to specialisations, from the greater efficiency of some major function like digestion to overall advance in general organisation; the third is stasigenesis or stabilised limitation of evolution. This occurs when specialisation for a particular way of life reaches a dead end as with horses, or efficiency of function attains a maximum as with hawks' vision, or an ancient type of organisation persists as a living fossil like the lungfish or the tuatara.

Major advance or biological progress is always by a succession of dominant types, each step achieved by a rare breakthrough from some established type of organisation to a new and more effective one, as from the amphibian to true dry-land reptilian type, or from the cold-blooded egg-laying reptile to the warm-blooded self-regulating mammal with intra-uterine development. The new dominant type multiplies at the expense of the old, which may become extinct (as did the jawless fish) or may persist in reduced numbers (as did the reptiles.)

The psychosocial phase, the latest of which we have any knowledge (though elsewhere in the universe there may have been a breakthrough to some new phase as unimaginable to us as the psychosocial phase would have been to even the most advanced Pliocene primate), is based on a self-reproducing and self-varying system of cumulative transmission of experience and culture, operating by mechanisms of psychological and social selection which we have not as yet adequately defined or analysed. Spatially it is very limited; we know of it only on this earth, and in any case it must be restricted to the surface of a small minority of planets in the small minority of stars possessing planetary systems. On our planet it is at the very beginning of its course, having begun less than one million years ago. However, its tempo is not only much faster than that of biological evolution, but manifests a new phenomenon, in the shape of a marked acceleration. Its overall trend is highly anti-entropic, and is characterised by a sharp increase in the operative significance of exceptional individuals and in the importance of true purpose and conscious evaluation based on reason and imagination, as against the automatic differential elimination of random variants.

The most significant element in that trend has been the growth and improved organisation of tested and established knowledge. And its results are psychologically (mentally) generated organizations even more astonishingly varied and complex than biological organisms – machines, concepts, cooking, mass communications, cities, philosophies, superstitions, propaganda, armies and navies, personalities, legal systems, works of art, political and economic systems, entertainments, slavery, scientific theories, hospitals, moral codes, prisons, myths, languages, torture, games and sports, religions, record and history, poetry, civil services, marriage systems, initiation rituals, agriculture, drama, social hierachies, schools and universities. Accordingly evolution in the human phase is no longer purely biological, based on changes in the material system of genetic transmission, but primarily cultural, based on changes in the psychosocial system of ideological and cultural transmission.

In the psychosocial phase of evolution the same three subprocesses operating in the biological phase are still at work – cladogenesis, operating to generate difference and variety within and between cultures; anagenesis, operating to produce improvement in detailed technological methods, in economic and political machinery, in administrative and educational systems, in scientific thinking and creative expression, in moral tone and religious attitude, in social and international organisation; and stasigenesis, operating to limit progress and to keep old systems and attitudes, including even outworn superstitions, persisting alongside of or actually within more advanced social and intellectual systems. But there is an additional fourth sub-process, that of convergence (or at least anti-divergence), operating by diffusion – diffusion of ideas and techniques between individuals, communities, cultures and regions. This is tending to give unity to the world: but we must see to it that it does not also impose uniformity and destroy desirable variety.

As in the biological phase, major advance in the human phase is brought about by a succession of generally or locally dominant types. These, however, are not types of organism, but of cultural and ideological organisation. Monotheism as against polytheism, for instance; or in the political sphere, the beginning of one-world internationalism as against competitive multinationalism. Or again, science as against magic, democracy as against tyranny, planning as against *laissez-faire*, tolerance as against intolerance, freedom of opinion and expression as against authoritarian dogma and repression.

Not only does the succession of dominant types bring about progress or advance in organisation within each of the three main evolutionary phases, but it also operates, though on a grander and more decisive scale, in the evolutionary process as a whole. A biological organism possesses a higher degree of organisation than any inorganic system; as soon as living organisms were produced, they became the major dominant type of organisation on earth, and the course of evolution became predominantly biological and only secondarily inorganic. Similarly a psychosocial system possesses a higher degree of organisation than any biological organism: accordingly man at once became the new major dominant type on earth, and the course of evolution became predominantly cultural and only secondarily biological, with inorganic methods quite subordinate.

The evolutionary perspective includes the broad background of the cosmic past. Now against this background we must face the problems of the present and the challenge of the future. Let me begin by reiterating that man is an exceedingly recent phenomenon. The earliest creatures which can properly be called men, though they must be assigned to different genera from ourselves, date from less than two million years ago, and our own species, *Homo sapiens*, from much less than half a million years. Man began to put his toe tentatively across the critical threshold leading towards evolutionary dominance perhaps a quarter of a million years ago, but took tens of thousands of years to cross it, emerging as a dominant type only during the last main glaciation, probably later than 100 000 BC, but not becoming fully dominant until the discovery of agriculture and stock-breeding well under 10 000 years ago, and overwhelmingly so with the invention of machines and writing and organised civilisation a bare five millennia before the present day, when his dominance has become so hubristic as to threaten his own future.

All new dominant types begin their career in a crude and imperfect form, which then needs drastic polishing and improvement before it can reveal its full potentialities and achieve full evolutionary success. Man is no exception to this rule. He is not merely exceedingly young; he is also exceedingly imperfect, an unfinished and often botched product of evolutionary improvisation. Teilhard de Chardin has called the transformation of an anthropoid into a man *hominisation*: it might be more accurately, though more clumsily, termed *psychosocialisation*. But whatever we call it, the process of hominisation is very far from complete, the serious study of its course, its mechanisms and its techniques has scarcely started, and only a fraction of its potential results have been realized. Man, in fact, is in urgent need of further improvement.

This is where eugenics comes into the picture. For though the psychosocial system in and by which man carries on his existence could obviously be enormously improved with great benefit to its members, the same is also true for his genetic outfit.

Severe and primarily genetic disabilities like haemophilia, colour-blindness, mongolism, some kinds of sexual deviation, much mental defect, sickle-cell anaemia, some forms of dwarfism, and Hunting-ton's chorea are the source of much individual distress, and their reduction would remove a considerable burden from suffering humanity. But these are occasional abnormalities. Quantitatively their total effect is insignificant in comparison with the massive imperfection of man as a species, and their reduction appears as a minor operation in comparison with the large-scale positive possibilities of all-around general improvement.

Take first the problem of intelligence. It is to man's higher level of intelligence that he owes his evolutionary dominance; and yet how low that level still remains! It is now well established that the human IQ, when properly assayed, is largely a measure of genetic endowment. Consider the difference in brain-power between the hordes of average men and women with IQ's around 100 and the meagre company of Terman's so-called geniuses with IQ's of 160 or over, and the much rarer true geniuses like Newton and Darwin, Tolstoy and Shakespeare, Goya and Michelangelo, Hammurabi and Confucius; and then reflect that, since the frequency curve for intelligence is approximately symmetrical, there are as many stupider people with IQ's below 100 as there are abler ones with IQ's above it.

Recollect also that the great and striking advances in human affairs, as much in creative art and political and military leadership as in scientific discovery and invention, are primarily due to a few exceptionally gifted individuals. Remember that on the established principles of genetics a small raising of average capacity would of necessity result in an upward shifting of the entire frequency curve, and therefore a considerable increase in the absolute numbers of such highly intelligent and well-endowed human beings that form the uppermost section of the curve (as well as a decrease in the numbers of highly stupid and feebly endowed individuals at the lower end).

Reflect further on the fact, originally pointed out by Galton, that there is already a shortage of brains capable of dealing with the complexities of modern administration, technology and planning, and that with the inevitable increase of our social and technical complexity, the greater will that shortage become. It is thus clear that for any major advance in national and international efficiency we cannot depend on haphazard tinkering with social or political symptoms or *ad hoc* patching up of the world's political machinery, or even on improving general education, but must rely increasingly on raising the genetic level of man's intellectual and practical abilities. As I shall later point out, artificial insemination by selected donors could bring about such a result in practice.

The same applies everywhere in the psychosocial process. For more and better scientists, we need the raising of the genetic level of exploratory curiosity or whatever it is that underlies single-minded investigation of the unknown and the discovery of novel facts and ideas; for more and better artists and writers, we need the raising of the genetic level for disciplined creative imagination; for more and better statesmen, that of the capacity to see social and political situations as wholes, to take long-term and total instead of only short-term and partial views; for more and better technologists and engineers, that of the passion and capacity to understand how things work and to make them work more efficiently; for more and better saints and moral leaders, that of disciplined valuation, of devotion and duty, and of the capacity of love; and for more and better leaders of thought and guides of action we need a raising of the capacity of man's vision and imagination, to form a comprehensive picture, at once reverent, assured and unafraid, of nature and man's relations with it.

These facts and ideas have an important bearing on the so-called race question and the problem of racial equality. I should rather say racial *inequality*, for up till quite recently the naïve belief in the natural inequality of races and people in general, and the inherent superiority of one's own race or people in particular, has almost universally prevailed.

To demonstrate the way in which this point of view permeated even nineteenth-century scientific thought, it is worth recalling that it was largely subscribed to by Darwin in his comments on the Fuegians in the *Voyage of the Beagle*, and in more general but more guarded terms in the *Descent of Man*: and Galton himself, against a similar background of travels among backward tribes and on the basis of his own rather curious method of assessment, concluded that different races had achieved different genetic standards, so that, for instance, 'the average standard of the Negro race is two grades below our own.' This type of belief, after being given a pseudo-scientific backing by non-biological theoreticians like Gobineau and Houston Stewart Chamberlain, was used to justify the Nazis' 'Nordic' claims to world domination and their horrible campaign for the extermination of the 'inferior, non-Aryan race' of Jews, and is still employed with support from Holy Writ and the majority of the Dutch Reformed Church in South Africa, to sanction Verwoerd's denial of full human rights to non-whites.

Later investigation has conclusively demonstrated first, that there is no such thing as a 'pure race.' Secondly, that the obvious differences in level of achievement between different peoples and ethnic groups are primarily cultural, due to differences not in genetic equipment but in historical and environmental opportunity. And thirdly, that when the potentialities of intelligence of two major 'races' or ethnic groups, such as whites and negroes or Europeans and Indians, are assessed as scientifically as possible, the frequency curves for the two groups overlap over almost the whole of their extent, so that approximately half the population of either group is genetically stupider (has a lower genetic IQ) than the genetically more intelligent half of the other. There are thus large differences in genetic mental endowment *within* single racial groups, but minimal ones *between* different racial groups.

Partly as a result of such studies, but also of the prevalent environmentalist views of Marxist and Western liberal thought, an anti-genetic view has recently developed about race. It is claimed that though ethnic groups obviously differ in physical characters, and that some of them, like pigmentation, nasal form, and number of sweat-glands, were originally adaptive, they do not (and sometimes even they cannot) differ in psychological or mental characters such as intelligence, imagination, or even temperament.*

Against this new pseudo-scientific racial naïveté, we must set the following scientific facts and principles. First, it is clear that the major human races originated as geographical sub-species of *Homo sapiens*, in adaptation to the very different environments to which they have become restricted. Later, of course, expansion and migration

---

* There is the further point that races may differ considerably in body-build and that Sheldon and others have made it highly probable that body-build is correlated with temperament. Unfortunately, racial differences in body-build have not yet been analysed in terms of Sheldon's somatotypes: here is an important field for research.

reversed the process of differentiation and led to an increasing degree of convergence by crossing, though considerable genetic differentiation remains. Secondly, as Professor Muller has pointed out, it is theoretically inconceivable that such marked physical differences as still persist between the main racial groups should not be accompanied by genetic differences in temperament and mental capacities, possibly of considerable extent. Finally, as previously explained, advance in cultural evolution is largely and increasingly dependent on exceptionally well-endowed individuals. Thus two racial groups might overlap over almost the whole range of genetic capacity, and yet one might be capable of considerably higher achievement, not merely because of better environmental and historical opportunity, but because it contained say 2 instead of 1 per cent of exceptionally gifted men and women. So far as I know, proper scientific research on this subject has never been carried out, and possibly our present methods of investigation are not adequate for doing so, but the principle is theoretically clear and is of vital practical importance.*

This does not imply any belief in crude racism, with its unscientific ascription of natural and permanent superiority or inferiority to entire races. As I have just pointed out, approximately half of any large ethnic group, however superior its self-image may be, is in point of fact genetically inferior to half of the rival ethnic group with which it happens to be in social or economic competition and which it too often stigmatises as permanently and inherently lower. Furthermore, practical experience demonstrates that every so-called race, however underdeveloped its economic and social system may happen to be, contains a vast reservoir of untapped talent, only waiting to be elicited by a combination of challenging opportunity, sound educational methods, and efficient special training. I recently attended an admirable symposium on nutrition in Nigeria where the scientific quality of the African contributions was every whit as high as that of the whites; and African politicians can be just as statesmanlike (and

---

* On the supposition that genetic intelligence is multifactorially (polygenically) determined and that its distribution follows a normal symmetrical curve, it can be calculated that the raising of the mean genetic IQ of a population by 11/2 per cent would result in a 50 per cent increase in the number of individuals with an IQ of 160 or over. The proportion of such highly-endowed individuals would rise from 1 in about 30 000 of the total population to 1 in about 20 000. Sir Cyril Burt informs me that if, as is possible, some types of high genetic intelligence are determined by single genes, the increase might be still greater.

also just as unscrupulously efficient in the political game) as their European or American counterparts.

The basic fact about the races of mankind is their almost total overlap in genetic potentialities. But the most significant fact for eugenic advance is the large difference in achievement made possible by a small increase in the number of exceptional individuals.

The evolutionary biologist can point out to the social scientist and the politician that this importance of the exceptional individual for psychosocial advance is merely an enhancement of a long-established evolutionary trend. Exceptional individuals can be important for biological improvement in mammals, in birds, and possibly even in insects. New food-traditions in Japanese monkeys are established by disobedient young individuals. The utilisation of milk-bottles as a new source of food by blue tits was due to the activities of a few exceptional tit geniuses in a few widely separate localities. All male satin bowerbirds construct bowers and assemble collections of stimulating bright objects at them, but only a minority deliberately paint their bowers with a mixture of berries, charcoal and saliva, and only a still smaller minority indulge the species' natural preference for blue objects by deliberately stealing bluebags to add to their display collection. And there is some evidence that even in ants, those prototypes of rigidly instinctive behaviour, a few workers are exceptionally well-endowed with the exploratory urge, and play a special role in discovering new sources of food for the colony.

But I must return to man as a species. The human species is in desperate need of genetic improvement if the whole process of psychosocial evolution which it has set in train is not to get bogged down in unplanned disorder, negated by over-multiplication, clogged up by mere complexity, or even blown to pieces by obsessional stupidity. Luckily it not only *must* but *can* be improved. It can be improved by the same type of method that has secured the improvement of life in general – from protozoan to protovertebrate, from protovertebrate to primate, from primate to human being – the method of multi-purpose selection directed towards greater achievement in the prevailing conditions of life.

On the other hand, it can *not* be improved by applying the methods of the professional stockbreeder. Indeed the whole discussion of eugenics has been bedevilled by the false analogy between artificial and natural selection. Artificial selection is intensive special-purpose selection, aimed at producing a particular excellence, whether in milk-yield in cattle, speed in race-horses or a fancy image in dogs. It produces a number of specialised pure breeds, each with a markedly lower variance than the parent species. Darwin rightly drew heavily on its results in order to demonstrate the efficacy of selection in general. But since he never occupied himself seriously with eugenics he did not point out the irrelevance of stock-breeding methods to human improvement. In fact, they are not only irrelevant, but would be disastrous. Man owes much of his evolutionary success to his unique variability. Any attempt to improve the human species must aim at retaining this fruitful diversity, while at the same time raising the level of excellence in all its desirable components, and remembering that the selectively evolved characters of organisms are always the results of compromise between different types of advantage, or between advantage and disadvantage.

Natural selection is something quite different. To start with, it is a shorthand metaphorical term coined by Darwin to denote the teleonomic or directive agencies affecting the process of evolution in nature, and operating through the differential survival and reproduction of genetical variants. It may operate between conspecific individuals, between conspecific populations, between related species, between higher taxa such as genera and families, or between still larger groups of different organisational type, such as Orders and Classes. It may also operate between predator and prey, between parasite and host, and between different synergic assemblages of species, such as symbiotic partnerships and ecological communities. It is in fact universal in its occurrence, though multiform in its mode of action.

Some over-enthusiastic geneticists appear to think that natural selection acts directly on the organism's genetic outfit or genotype. This is not so. Natural selection exerts its effects on animals and plants as working mechanisms: it can operate only on phenotypes. Its evolutionary action in transforming the genetic outfit of a species is indirect, and depends on the simple fact pointed out by Darwin in the *Origin* that much variation is heritable – in modern terms, that there is a high degree of correlation between phenotypic and genotypic variance. The correlation, however, is never complete, and there are many cases where it is impossible without experimental analysis to determine whether a variant is modificational, due to alteration in the environment, or mutational, due to alteration in the genetic outfit. In certain cases, environmental treatment will produce so-called phenocopies which are indistinguishable from mutants in their visible appearance.

This last fact has led Waddington to an important discovery – the fact that an apparently Lamarckian mode of evolutionary transformation can be precisely simulated by what he calls genetic assimilation. To take an actual example, the rearing of fruitfly larvae on highly saline media produces a hypertrophy of their salt-excreting glands through direct modification. But if selection is practised by breeding from those individuals which show the maximum hypertrophy of their glands, then after some ten or twelve generations, individuals with somewhat hypertrophied glands appear even in cultures on non-saline media. The species has a genetic predisposition, doubtless brought by selection in the past, to react to saline conditions by glandular hypertrophy. The action of the major genes concerned in reactions of this sort can be enhanced (or inhibited) by so-called modifying genes of minor effect. Selection has simply amassed in the genetic outfit an array of such minor enhancing genes strong enough to produce glandular hypertrophy even in the absence of any environmental stimulus. It is only pseudo-Lamarckism, but no less important for that – a significant addition to the theoretical armoury of evolutionary science.

I repeat that the most important effect achieved by natural selection is biological improvement. As G. G. Simpson reminds us, it does so opportunistically, making use of whatever new combination of existing mutant genes, or less frequently of whatever new mutations, happens to confer differential survival value on its possessors. We know of numerous cases where phenotypically identical and adaptive transformations have been produced by different genes or gene-combinations.

Here I must digress a moment to discuss the concept of evolutionary fitness. The biological *avant garde* has chosen to define *fitness* as 'net reproductive advantage,' to use the actual words employed by Professor Medawar in his Reith Lectures on *The Future of Man*. Any strain of animal, plant or man which leaves slightly more descendants capable of reproducing themselves than another, is then defined as 'fitter.' This I believe to be an unscientific and misleading definition. It disregards all scientific conventions as to priority, for it bears no resemblance to what Spencer implied or intended by this famous phrase the *survival of the fittest*.* It is also nonsensical in every context save the limited field of population genetics. In biology,

---

* Darwin did not use the phrase in the first edition of the *Origin of Species*, though in later editions he added it as an equivalent to natural selection – see footnote on page 1 of this book.

fitness must be defined, as Darwin did with improvement, 'in relation to the conditions of life' – in other words, in the context of the general evolutionary situation. I shall call it *evolutionary fitness*, in contradistinction to the purely reproductive fitness of the evangelists of geneticism, which I prefer to designate by the descriptive label of *net* or *differential reproductive advantage*.

Meanwhile, I have a strong suspicion that the genetical *avant garde* of today will become the rearguard of tomorrow. In my own active career I have seen a reversal of this sort in relation to natural selection and adaptation. During the first two decades of this century the biological *avant garde* dismissed topics such as cryptic or mimetic coloration, and indeed most discussion of adaptation, as mere 'armchair speculation,' and played down the role of natural selection in evolution, as against that of large and random mutation. Bateson's enthusiasm rebounded from his early protest against speculative phylogeny into the far more speculative suggestion made *ex cathedra* at a British Association meeting, that all evolution, whether of higher from lower, or of diversity from uniformity, had been brought about by loss mutations; and the great T. H. Morgan once permitted himself to state in print that, if natural selection had never operated, we should possess all the organisms that now exist and a great number of other types as well! This anti-selectionist *avant garde* of fifty years back has now come over *en masse* into the selectionist camp, leaving only a few retreating stragglers to deliver some rather ineffective parthian shots at their opponents.

Natural selection is a teleonomic or directional agency. It utilises the inherent genetic variability of organisms provided by the raw material of random mutation and chance recombination, and it operates by the simple mechanism of differential reproductive advantage. But on the evolutionary time-scale it produces biological improvement, resulting in a higher total and especially a higher upper level of evolutionary fitness, involving greater functional efficiency, higher degrees of organisation, more effective adaptation, better self-regulating capacity, and finally more mind – in other words an enrichment of qualitative awareness coupled with more flexible behaviour.

Man almost certainly has the largest reservoir of genetical variance of any natural species: selection for the differential reproduction of desirable permutations and combinations of the elements of this huge variance could undoubtedly bring about radical improvement in the human organism, just as it has in pre-human types. But the agency of human transformation cannot be the blind and automatic natural selection of the pre-human sector. That, as I have already stressed, has been relegated to a subsidiary role in the human phase of evolution. Some form of psychosocial selection is needed, a selection as non-natural as are most human activities, such as wearing clothes, going to war, cooking food, or employing arbitrary systems of communication. To be effective, such 'non-natural' selection must be conscious, purposeful and planned. And since the tempo of cultural evolution is many thousands of times faster than that of biological transformation, it must operate at a far higher speed than natural selection if it is to prevent disaster, let alone produce improvement.

Luckily there is today at least the possibility of meeting both these prerequisites: we now possess an accumulation of established knowledge and an array of tested methods which could make intelligent, scientific and purposeful planning possible. And we are in the process of discovering new techniques which could raise the effective speed of the selective process to a new order of magnitude. The relevant new knowledge mainly concerns the various aspects of the evolutionary process – the fact that there are no absolutes or all-or-nothing effects in evolution and that all organisms and all their phenotypic characters represent a compromise or balance between competing advantages and disadvantages; the effect of selection on populations in different environmental conditions; the origin of adaptation; and the general improvement of different evolutionary lines in relation to the conditions of their life. The notable new techniques include effective methods of birth-control, the successful development of grafted fertilised ova in new host-mothers, artificial insemination, and the conservation of function in deep-frozen gametes.

We must first keep in mind the elementary but often neglected fact that the characters of organisms which make for evolutionary success or failure, are not inherited as such. On the contrary, they develop anew in each individual, and are always the resultant of an interaction between genetic determination and environmental modification. Biologists are often asked whether heredity or environment is the more important. It cannot be too often emphasised that the question should never be asked. It is as logically improper to ask a biologist to answer it as it is for a prosecuting counsel to ask a defendant when he stopped beating his wife. It is the phenotype which is biologically significant and the phenotype is a resultant produced by the complex interaction of hereditary and environmental factors. Eugenics, in common with evolutionary biology in general, needs this phenotypic approach.

Man's evolution occurs on two different levels and by two distinct methods, the genetic, based on the transmission and variation of genes and gene-combinations, and the psychosocial or cultural, based on the transmission and variation of knowledge and ideas.

Professor Medawar, in his Reith Lectures on *The Future of Man*, while admitting in his final chapter that man possesses 'a new, non-genetical, system of heredity and evolution' (p. 88), claims on p. 41 that this is 'a new kind of biological evolution (I emphasise, a biological evolution).' I must insist that this is incorrect. The psychosocial process – in other words, evolving man – is a new *state* of evolution, a new *phase* of the cosmic process, as radically different from the pre-human biological phase as that is from the inorganic or pre-biological phase; and this fact has important implications for eugenics.

An equally elementary but again often neglected fact is that organisms are not significant – in plain words, are meaningless – except in relation to their environment. A fish is not a thing-in-itself: it is a type of organism evolved in relation to an active life in large or medium-sized bodies of water. A cactus has biological significance only in relation to an arid habitat, a woodpecker only in relation to an arboreal one. Man, however, is in a unique situation. He must live not only in relation with the physicochemical and biological environment provided by nature, but with the psychosocial environment of material and mental habitats which he has himself created.

Man's psychosocial environment includes his beliefs and purposes, his ideals and his aims: these are concerned with what we may call the habitat of the future, and help to determine the *direction* of his further evolution. All evolution is directional and therefore relative. But whereas the direction of biological evolution is related to the continuing improvement of organisms in relation to their conditions of life, human evolution is related to the improvement of the entire psychosocial process, including the human organism, in relation to man's purposes and beliefs, long-term as well as short-term. Only in so far as those purposes and beliefs are grounded on scientific and tested knowledge, will they serve to steer human evolution in a desirable direction. In brief, biological evolution is given direction by the blind and automatic agency of natural selection operating through material mechanisms, human evolution by the agency of psychosocial guidance operating with the aid of mental awareness, notably the mechanisms of reason and imagination.

To be effective, such awareness must clearly be concerned with man's environmental situation as well as his genetic equipment. In my first Galton Lecture, I pointed out the desirability of eugenists relating their policies to the social environment. Today I would go further, and stress the need for planning the environment in such a way as will promote our eugenic aims. By 1936, it was already clear that the net effect of present-day social policies could not be eugenic, and was in all probability dysgenic. But, as Muller has demonstrated, this was not always so. In that long period of human history during which our evolving and expanding hominid ancestors lived in small and tightly knit groups competing for territorial and technological success, the social organisation promoted selection for intelligent exploration of possibilities, devotion and co-operative altruism: the cultural and the genetic systems reinforced each other. It was only much later, with the growth of bigger social units of highly organised civilisations based on status and class differentials, that the two became antagonistic; the sign of genetic transformation changed from positive to negative and definite genetic improvement and advance began to halt, and gave way to the possibility and later the probability of genetic regression and degeneration.

This probability has been very much heightened during the last century, partly by the differential multiplication of economically less favoured classes and groups in many parts of the world, partly by the progress of medicine and public health, which has permitted numbers of genetically defective human beings to survive and reproduce; and today it has been converted into a certainty by the series of atomic and nuclear explosions which have been set off since the end of the last war. There is still dispute as to the degree of damage this has done to man's genetic equipment. There can be no dispute as to the fact of damage: any addition to man's load of mutations can only be deleterious, even if some of them may possibly come to be utilised in neutral or even favourable new genecombinations.

Now that we have realised these portentous facts, it is up to us to reverse the process and to plan a society which will favour the increase instead of the decrease of man's desirable genetic capacities for intelligence and imagination, empathy and co-operation, and a sense of discipline and duty.

The first step must be to frame and put into operation a policy designed to reduce the rate of human increase before the quantitative claims of mere numbers override those of quality and prevent any real improvement, social and economic as much as eugenic. I would prophesy that within a quite short time, historically speaking, we shall find ourselves aiming at an absolute reduction of the population in the world in general, and in overcrowded countries like Britain, India and China, Japan, Java and Jamaica in particular; the quantitative control of population is a necessary prerequisite for qualitative improvement, whether psychosocial or genetic.

Science seems to be nearing a breakthrough to cheap and simple methods of birth-control, or reproduction-control as it should more properly be called. The immediate needs are for much-increased finance for research, testing, pilot projects, motivation studies and the education of public opinion, and an organised campaign against the irrational attitudes and illiberal policies of various religious and political organisations. Simultaneously, responsible opinion must begin to think out ways in which social and economic measures can be made to promote desirable genetic trends and reproductive habits.

Many countries have instituted family allowance systems which are not graded according to number of children, and some, like France, even provide financial inducements which encourage undesirably large families. It should be easy to devise graded family allowance systems in which the allowances for the first two or three children would be really generous, but those for further children would rapidly taper off. In India, there have even been proposals to tax parents for children above a certain number, and in some provinces, men fulfilling certain conditions are paid to be vasectomised.

A powerful weapon for adequate population-control is ready to the hand of the great grant-giving and aid-providing agencies of the modern world – international agencies such as the UN and its Technical Assistance Board representing its various Specialised Agencies like FAO and UNESCO, the World Bank and the International Finance Corporation Administration; national agencies like the Colombo Plan and the Inter-American Development Fund; and the great private Foundations (wittily categorised as *philanthropoid* by that remarkable man Frederick Keppel) like Rockefeller and Ford, Gulbenkian, Nuffield and Carnegie.

At the moment, much of the financial and technical aid provided by these admirable bodies is being wasted by being flushed down the drain of excess population instead of into the channels of positive economic and cultural development, or is even defeating its own ends by promoting excessive and over-rapid population-increase.

Bankers do not make loans unless they are satisfied of the borrower's credit-worthiness. Surely these powerful agencies, public or private, should not provide loans or grants or other aid unless they are satisfied of the recipient nation's demographic credit-worthiness. If an under-developed nation's birth-rate is excessive, the aid will go in providing the basic minima of food, care, shelter and education for the flood of babies, instead of the capital and the technical skills needed to achieve the breakthrough to a viable industrialisation. Wherever this is so, the aid-providing institution should insist that the nation should frame an approved policy of population-control, and that some of the aid should be devoted to the implementation of that policy and to research on the subject. And the UN should, of course, take steps to prepare the way for a World Population Policy, should carry out or in any case encourage research on population-control, and should ensure that its Specialized Agencies like WHO, UNESCO, FAO and ILO, pay due attention to the problems of population in relation to their special fields of competence.

At last I reach my specific subject – eugenics, with its two aspects, negative and positive. Negative eugenics aims at preventing the spread and especially the increase of defective or undesirable human genes or gene-combinations, positive eugenics at securing the reproduction and especially the increase of favourable or desirable ones.*

Negative eugenics has become increasingly urgent with the increase of mutations due to atomic fallout, and with the increased survival of genetically defective human beings, brought about by advances in medicine, public health, and social welfare. But it must, of course, attempt to reduce the incidence, or the manifestation, of every kind of genetic defect. Such defects include high genetic proneness to diseases such as diabetes, schizophrenia (which affects 1 per cent of the entire human population), other insanities, myopia, mental defect and very low IQ, as well as more clear-cut defects like colour-blindness or haemophilia.

When defects depend on a single dominant gene, as with Huntington's chorea, transmission can of course be readily prevented by persuading the patient to refrain from reproducing himself. With sexlinked defects like haemophilia, Duchenne-type muscular dystrophy, or HCN 'smell-blindness,' this will help, but the method

---

* In the past, these aims have been generally expressed in terms of defective or desirable *stocks* or *strains*. With the progress of genetics, it is better to reformulate them in terms of genes.

should be supplemented by counselling his sisters against marriage. This will be more effective and more acceptable when, as seems possible, we can distinguish carriers heterozygous for the defect from non-carriers. This is already practicable with some autosomic recessive defects, notably sickle-cell anaemia. Here, registers of carriers have been established in some regions, and they are being effectively advised against intermarriage. This will at least prevent the manifestation of the defect. The same could happen with galactosaemia, and might be applicable to relatives of patients with defects like phenylketonuria and agammoglobulinaemia.

In addition, the marked differential increase of lower-income groups, classes and communities during the last hundred years cannot possibly be eugenic in its effects. The extremely high fertility of the so-called problem group in the slums of many industrial cities is certainly anti-eugenic.

As Muller and others have emphasised, unless these trends can be checked or reversed, the human species is threatened with genetic deterioration, and unless this load of defects is reduced, positive eugenics cannot be successfully implemented. For this we must reduce the reproduction rate of genetically defective individuals: that is negative eugenics.

The implementation of negative eugenics can only be successful if family planning and eugenic aims are incorporated into medicine in general and into public health and other social services in particular. Its implementation in practice will depend on the use of methods of contraception or sterilisation, combined where possible with AID (artificial insemination by donor) or other methods of vicarious parenthood. In any case, negative eugenics is of minor evolutionary importance and the need for it will gradually be superseded by efficient measures of positive eugenics.

In cases of specific genetic defect, voluntary sterilisation is probably the best answer.* In the defective married male, it should be coupled with artificial parenthood (AP) by donor insemination (AID) as the source of children. In the defective female, the fulfilments of child-rearing and family life will have to be secured by adoption until such time – which may not be very distant – as improved technique makes possible artificial parenthood by transfer of fertilized ova, which we may call AOD. In both cases, it must be

---

* It will be even more satisfactory if, as now appears likely, reversible male sterilisation (vasectomy) becomes practicable.

remembered that sterilization does not prevent normal healthy and happy sexual intercourse.

Certified patients are now prevented from reproducing themselves by being confined in mental hospitals. If sterilised, they might be allowed to marry if this were considered likely to ameliorate their condition.

In the case of the so-called social problem group, somewhat different methods will be needed. By social problem group I mean the people, all too familiar to social workers in large cities, who seem to have ceased to care, and just carry on the business of bare existence in the midst of extreme poverty and squalor. All too frequently they have to be supported out of public funds, and become a burden on the community. Unfortunately they are not deterred by the conditions of existence from carrying on with the business of reproduction: and their mean family size is very high, much higher than the average for the whole country.

Intelligence and other tests have revealed that they have a very low average IQ; and the indications are that they are genetically subnormal in many other qualities, such as initiative, pertinacity, general exploratory urge and interest, energy, emotional intensity, and willpower. In the main, their misery and improvidence is not their fault but their misfortune. Our social system provides the soil on which they can grow and multiply, but with no prospects save povery and squalor.

Here again, voluntary sterilisation could be useful. But our best hope, I think, must lie in the perfection of new, simple and acceptable methods of birth-control, whether by an oral contraceptive or perhaps preferably by immunological methods involving injections. Compulsory or semi-compulsory vaccination, inoculation and isolation are used in respect of many public health risks: I see no reason why similar measures should not be used in respect of this grave problem, grave both for society and for the unfortunate people whose increase has been actually encouraged by our social system.

Many social scientists and social workers in the West, as well as all orthodox Marxists, are environmentalists. They seem to believe that all or most human defects, including many that western biologists would regard as genetic, can be dealt with, cured or prevented by improving social environment and social organisation. Even some biologists, like Professor Medawar, agree in general with this view, though he admits a limited role for negative eugenics, in the shape of what he calls 'genetic engineering.' For him, the 'newer solution' of the problem, which 'goes some way towards making up for the inborn inequalities of man,' is simply to improve the environment. With this I cannot agree. Although certain particular problems can be dealt with in this way, for instance proneness to tuberculosis by improving living conditions and preventing infection, such methods cannot cope with the general problem of genetic deterioration, because this, if not checked, will steadily increase through the accumulation of mutant genes which otherwise would have been eliminated.

It is true that many diseases or defects with a genetic basis, like diabetes or myopia, can be cured by treatment, though almost always with some expense, trouble, or dicomfort to the defective person as well as to society. But if the incidence of such defects (not to mention the many others for which no cure or remedy is now known) were progressively multiplied, the burden would grow heavier and heavier and eventually wreck the social system. As in all other fields, we need to combine environmental and genetic measures, and if possible render them mutually reinforcing.

Against the threat of genetic deterioration through nuclear fall-out there are only two courses open. One is to ban all nuclear weapons and stop bomb-testing; the other is to take avantage of the fact that deep-frozen mammalian sperm will survive, with its fertilising and genetic properties unimpaired, for a long period of time and perhaps indefinitely, and accordingly to build deep shelters for sperm-banks – collections of deep-frozen sperm from a representative sample of healthy and intelligent males. A complete answer must wait for the successful deep-freezing of ova also. But this may be achieved in the fairly near future, and in any case shelters for sperm-banks will give better genetic results than shelters for people, as well as being very much cheaper.

Positive eugenics has a far larger scope and importance than negative. It is not concerned merely to prevent genetic deterioration, but aims to raise human capacity and performance to a new level.

For this, however, it cannot rely on measures designed to produce merely a slight differential increase of genetically superior stocks, generation by generation. This is the way natural selection obtains its results, and it worked all right during the biological phase, when immense spans of time were available. But with the accelerated tempo of modern psychosocial evolution, much quicker results are essential. Luckily modern science is providing the necessary techniques, in the shape of artificial insemination and the deep-freezing of human gametes. The effects of superior germ-plasm can be multi-

plied ten or a hundredfold through the use of what I call EID –
eugenic insemination by deliberately preferred donors – and many
thousandfold if the superior sperm is deep-frozen.

This multiplicative method, harnessing man's deep desires for a
better future, was first put forward by H. J. Muller and elaborated by
Herbert Brewer, who invented the terms *eutelegenesis* and *agapogeny*
for different aspects of it. Some such method, or what we may term
Euselection – deliberate encouragement of superior genetic endowment – will produce immediate results. Couples who adopt this
method of vicarious parenthood will be rewarded by children outstanding in qualities admired and preferred by the couple themselves.

When deep-frozen ova too can be successfully engrafted into
women, the speed and efficiency of the process could of course be
intensified.

Various critics insist on the need for far more detailed knowledge
of genetics and selection before we can frame a satisfactory eugenic
policy or even reach an understanding of evolution. I can only say
how grateful I am that neither Galton nor Darwin shared these views,
and state my own firm belief that they are not valid. Darwin knew
nothing – I repeat *nothing* – about the actual mechanism of biological
variation and inheritance: yet he was possessed of what I can only call
a common-sense genius which gave him a general understanding of
the biological process and enabled him to frame a theory of the
process whose core remains unshaken and which has been able
successfully to incorporate all the modifications and refinements of
recent field study and genetic experiment.

Neither did the automatic process of natural selection 'know'
anything about the mechanisms of evolution. Luckily this did not
prevent it from achieving a staggering degree of evolutionary transformation, including miracles of adaptation and improvement. From
his seminal idea, Darwin was able to deduce important general
principles, notably that natural selection would automatically tend to
produce both diversification (adaptive radiation) and improvement
(biological advance or progress) in organisation, but that lower types
of organisation would inevitably survive alongside higher.

Critics of positive eugenics like Medawar inveigh against what they
call '*geneticism*.' However, he himself is guilty on this count, for he
has swallowed the population geneticists' claim (which I have discussed earlier) that theirs is the only scientifically valid definition of
fitness; and this is in spite of his admission that one organic type can
be more 'advanced' than another, and that 'human beings are the

outcome of a process which can perfectly well be described as an advancement.' However, he equates advancement with mere increase in complexity of the 'genetical instructions' given to the animal: if he had thought in broad evolutionary instead of restricted genetic terms he would have seen that biological advance involves improved organisation of the phenotype; that fitness in the geneticismal sense is a purely reproductive fitness; and that we must also take into account immediate phenotypic fitness and long-term evolutionary fitness. To put it in a slightly different way, 'fitness' as measured by differential survival of offspring is merely the mechanism by which the long-term improvement of true biological fitness is realised.

Recent genetic studies have shown the wide-spread occurrence of genetic polymorphism, in animal species and man, whether in the form of sharply distinct morphs (as with colour-blindness and other sensory morphisms), in multiplicity of slightly different alleles, or merely in a very high degree of potential variance. Some critics of positive eugenics maintain that this state of affairs will prevent or at least strongly impede any large-scale genetic improvement, owing to the resistance to change offered by genetic polymorphisms maintained by means of heterozygote advantage, which appear to comprise the majority of polymorphic systems.

It has further been suggested, notably by Professor Penrose, that people heterozygous for genes determining general intellectual ability, and therefore of medium or mediocre intelligence, are reproductively 'fitter' – more fertile – than those of high or low intelligence, and accordingly that, as regards genetic intelligence, the British population is in a state of natural balance. If so, it would be difficult to try to raise its average level by deliberate selective measures, and equally difficult for the level to sink automatically as the result of differential fertility of the less intelligent groups.

Although Medawar (op. cit. p. 125) appears to disagree with Penrose's main contention, he concludes that: 'If a tyrant were to attempt to raise the intelligence of all of us to its present maximum . . . I feel sure that his efforts would be self-defeating: the population would dwindle in numbers and, in the extreme case, might die out.' It is true that he later enters a number of minor caveats, but his main conclusion remains. This to me appears incomprehensible. If selection has operated, as it certainly has done in the past, during the passage from Pithecanthropus to present-day man, to bring about a very large rise in the level of genetic intelligence, why can it not bring about a much smaller rise in the immediate future? There are no grounds for believing that modern man's system of genetic variance differs significantly from that of his early human ancestors.

As regards balanced morphisms, it is of course true that they constitute stable elements in an organism's genotype. However, when their stability is mainly due to linkage with a lethal, and therefore to double-dose disadvantage rather than to heterozygote advantage, they may be destabilized by breaking the linkage. In any case, morphisms stable in one environment may sometimes be broken up in another. This has happened, for instance, with the white-yellow sex-limited morphism of the butterfly *Colias eurythema*, which in high latitudes has ceased to exist, and the local population is monomorphic, all homozygous white.

Certainly some morphisms show very high stability. For instance the PTC (phenylthiocarbamide) taste morphism occurs in apparently identical form both in chimpanzees and man, and so must presumably have resisted change for something like 10 million years. However, this remarkable stability of a specific genotypic component of the primate stock has not prevented the transformation of one branch of that stock into man!

Similar arguments apply to linked polygenic systems and to the general heterozygosity in respect of small allelic differences shown by so many organisms, including man. In the former case, Mather has shown how selection can break the linkage and make the frozen variability available for new recombinations and new evolutionary change. In the latter case, the stability need not be so intense as with clear-cut morphisms.

Frequently, it appears, polymorphism depends not so much on heterotic advantage as on a varying balance of advantage between the alleles concerned in different conditions: one allele is more advantageous in certain conditions, another in other conditions. The polymorphism is therefore a form of insurance against extreme external changes and gives flexibility in a cyclically or irregularly varying environment (Huxley, 1955). Such loose polymorphic systems can readily be modified by the incorporation of new and the elimination of old mutant alleles and the incorporation of new ones in response to directional changes in environment. In any case, their widespread existence has not stood in the way of directional evolutionary change, including the transformation of a protohominid into man. Why should they stand in the way of man's further genetic evolution?

The same reasoning applies to those numerous cases where high genetic variance, actual or potential, is brought about by multiple genic polymorphism, when many genes of similar action exist, often in a number of slightly different allelic forms.

In all these cases the critics of eugenics have been guilty of that very 'geneticism' which they deplore. They approach the subject from the standpoint of population genetics. If they were to look at it from an evolutionary standpoint, their difficulties would evaporate, and they would see that their objections could not be maintained.

Two further objections are often made to positive eugenics. One is by way of a question – who is to decide which type to select for? The other, which is by way of an answer to the first, is to assert that effective selection needs authoritarian methods and can only be put into operation by some form of dogmatic tyranny, usually stigmatised as intolerable or odious.

Both these objections reveal the same lack of understanding of psychosocial evolution as the genetical objections revealed about biological evolution: more simply, they demonstrate the same lack of faith in the potentialities of man that the purely genetical objections showed in the actual operative realisations of life.

For one thing, dogmatic tyranny in the modern world is becoming increasingly self-defeating: partly because it is dogmatic and therefore essentially unscientific, partly because it is tyrannical and therefore in the long run intolerable. But the chief point is that human improvement never works solely or even mainly by such methods and is doing so less and less as man commits himself more thoroughly to the process of general self-education.

Let me take an example. Birth-control resembles eugenics in being concerned with that most violent arouser of emotion and prejudice, human reproduction. However, during my own career, I have witnessed the subject break out of the dark prison of taboo into the international limelight. It was only in 1917 that Margaret Sanger was given a jail sentence for disseminating birth-control information. In the late twenties, when I was already over forty, I was summoned before the first Director-General of the BBC, now Lord Reith, and rebuked for having contaminated the British ether with such a shocking subject. Yet two years ago an international gathering in New York paid tribute to Margaret Sanger as one of the great women of our age. *Time* and *Life* Magazines both published long and reasoned articles on how to deal with the population explosion, and two official US committees reported in favour of the US conducting more research on birth-control methods and even of giving advice on the subject if requested by other nations. And today one can hardly open a copy of the most respectable newspapers without finding at least one reference to the grievous effects of population increase and population density on one or another aspect of human life in one or another country of the globe, including our own. Meanwhile, six nations have started official policies of family planning and population control, and many others are unofficially encouraging them.

Birth-control, in fact, has broken through – and in so doing it has changed its character and its methods. It began as a humanitarian campaign for the relief of suffering human womanhood, conducted by a handful of heroic figures, mostly women. It has now become an important social, economic and political campaign, led by powerful private associations, and sometimes the official or semi-official concern of national governments. Truth, in fact, prevails – though its prevailing demands time, public opprobrium of the self-sacrificing pioneers at the outset, and public discussion, backed by massive dissemination of facts and ideas, to follow.

We can safely envisage the same sort of sequence for evolutionary eugenics, operating by what may be called Euselection, though doubtless with much difference in detail. Thus the time to achieve public breakthrough might be longer because the idea of Euselection by delegated paternity runs counter to a deep-rooted sense of proprietary parenthood. On the other hand it might be shorter, since there is such a rapid increase in the popular understanding of science and in the agencies of mass communication and information, and above all because of the profound dissatisfaction with traditional ideas and social systems, which portends the drastic recasting of thought and attitude that I call the Humanist Revolution.

Some things, at least, are clear. First, we need to establish the legality, the respectability, and indeed the morality of AID. It must be cleared of the stigma of sin ascribed to it by Church dignitaries like Lord Fisher when Archbishop of Canterbury, and from the legal difficulties to its practice raised by the lawyers and administrators. Most importantly, the notion of donor secrecy must be abolished. Parents desiring AID should have not only the right but the duty of choice. For the time being, it may possibly be best that the name and personal identity of a donor should not be known to the acceptors, but there should certainly be a register of certified donors kept by medical men (and I would hope by the National Health Service) which would give particulars of their family histories. This would enable acceptors to exert a degree of conscious selection in choosing the father of the child they desire, and so pave the way for the supersession of blind and secrecy-ridden AID by an open-eyed and proudly accepted EID where the E stands for *Eugenic*.

The pioneers of EID, whether its publicists or its practitioners, will undoubtedly suffer all kinds of abusive prejudice – they will be accused of mortal sin, of theological impropriety, of immoral and unnatural practices. But they can take heart from what has happened in the field of birth-control, and can be confident that the rational control of reproduction aimed at the prevention of human suffering and frustration and the promotion of human well-being and fulfilment will in the not too distant future come to be recognised as a moral imperative.

The answers to the questions I mentioned at the beginning of this section are now, I hope, clear. There will be no single type to be selected for, but a range of preferred types; and this will not be chosen by any single individual or committee. The choice will be a collective choice representing the varied preferences and ideals of all the couples practising euselection by EID, and it will not be dogmatically imposed by any authoritarian agency, though as general acceptance of the method grows, it will be reinforced by public opinion and official leadership. The way is open for the most significant step in the progress of mankind – the deliberate improvement of the species by scientific and democratic methods.

All the objections of principle to a policy of positive eugenics fall to the ground when the subject is looked at in the embracing perspective of evolution, instead of in the limited perspective of population genetics or the short-term perspective of existing socio-political organization. Meanwhile the obvious practical difficulties in the way of its execution are being surmounted, or at least rendered surmountable, by scientific discovery and technical advance.

In evolutionary perspective, eugenics – the progressive genetic improvement of the human species – inevitably takes its place among the major aims of evolving man. What should we eugenists do in the short term to promote this long-term aim? We must of course continue to do and to encourage research on human genetics and reproduction, including methods of conception-control and sterilisation. The establishment of the Darwin Research Fellowships is an important milestone in this field: I hope that we shall be able to enlarge our research activities in the future.

We must continue to support negative eugenic measures, especially perhaps in respect of the so-called Social Problem group. We should assuredly continue to be concerned about population increase, and to support all agencies and organisations aiming at sane and scientific policies of population-control. We must equally support all agencies giving eugenic advice and marriage guidance. Since significant eugenic improvement depends on donor insemination, we must do all we can to win public support for AID, and to improve current practices in the subject.

In general, we must bring home to the general public the possibility of real genetic improvement, the burden it could lift off human shoulders, the hope it could kindle in human hearts. We must make people understand that social and cultural amelioration are not enough. If they are not to turn into temporary palliatives or degenerate into mere environmental tinkering, they must be combined with genetic amelioration, or at least with the hope of it in the future.

To ensure this, not only must the eugenics movement help to educate the public and especially the members of the professions – medical, educational, scientific, administrative, and others – in respect of eugenics, but it must make every effort to get the educational system improved at all levels, so as to provide everyone with the necessary minimum of biological understanding – an understanding of reproduction and population, genetics and selection, ecology and conservation, and above all of the process of evolution in its awe-inspiring sweep and of man's specific significance and responsibility in that comprehensive process.

If, as I firmly believe, man's role is to do the best he can to manage the evolutionary process on this planet and to guide its future course in a desirable direction, fuller realisation of genetic possibilities becomes a major motivation for man's efforts, and eugenics is revealed as one of the basic human sciences.

References

Baker, P. T. 1960. Climate, Culture and Evolution. *Human Biol.*, 32, 3 (Race, environment and culture.)

Blacker, C. P. 1952. *Eugenics: Galton and After*. Duckworth. London. (Galton's work and views. Modern developments in eugenics.)

Boyd, W. C. 1950 *Genetics and the Races of Man*. Little, Brown. Boston (Race and evolution.)

Brewer, H. 1935. Eutelegenesis: *Eugen. Rev.*, 27, 121.

—— 1939. Eutelegenesis. *Lancet*. 1939. 1, 265.

—— 1961. Ethical Parenthood and Contraception. *Balanced Living*. Brookville, Ohio. (17)3, p. 69.

Burt, C. 1958. The Inheritance of Mental Ability. *Amer. Psychol.* 13, 1. (Eugenics and intelligence.)

—— 1959, 1961. Class Differences in General Intelligence. *Brit. J. statist. Psychol*, 12, 15.

—— Intelligence and social mobility. *Ibid.*, 14, 3. (Distribution of intelligence.)

—— 1962. The Gifted Child. *Yearbook of Education*, 1962, p. 1. (Genetic intelligence.)

Carter, C. F. 1961. The Economic Use of Brains. *Advanc. Sci., Lond.*, 18, 222. (Shortage of individuals of high ability.)

Coale, A. J. and Hoover, E. M. 1959. *Population Growth and Economic Development in Low Income Countries*. (O.U.P., Bombay.)

Count, E. W. 1950. *This is Race*. Schuman. New York. (History of racial concepts.)

Darlington, C. D. 1960. (Review of P. B. Medawar's *The Future of Man*.) *Heredity*, 15, 44.

Dobzhansky, Th. 1962. *Mankind Evolving*. Yale Univ. Press. New Haven and London. D.N.A. (Population genetics, race, polymorphism, fitness, eugenics, cultural evolution, genotype and phenotype.)

Fisher, R. A., Ford, E. B. and Huxley, J. S. 1939. Taste-testing the anthropoid apes. *Nature*, 144, 750. (Sensory morphisms, apes and man.)

Ford, E. B., 1949. Polymorphism. *Biol. Rev.*, 20, 73. (Morphism, blood groups and disease.)

—— 1956. *Genetics for Medical Students*. Methuen. London (Human genetics; morphism.)

Hulse, F. S. 1960. Adaptation, Selection and Plasticity in Ongoing Human Evolution. *Human Biol.*, 32, 63. (Genetic plasticity and environment.)

Huxley, J. S. 1953. *Evolution in Action*. London and New York. Chatto and Windus (Natural selection and biological improvement.)

—— 1955. Morphism and Evolution, *Heredity*, 9, 1.

Huxley, J. S. (Ed.): 1961. *The Humanist Frame*. Allen and Unwin and Harpers. London and New York. (Introductory Chapter. General evolutionary theory.)

Kalmus, H. 1959. Genetical Variation and Sense-Perception. *Ciba Found. Symp., Biochem. Human Genetics*, 60. (Sensory defects and morphisms.)

Livingstone, F. B. 1960. Natural Selection, Disease and Ongoing Human Evolution. *Human Biol.*, 32, 17.

Mather, K. 1953. Genetical Control of Stability in Development. *Heredity*, 7, 297. (Balanced polymorphism, stability, release of variance.)

—— 1956. Polygenic Mutation and Variation in Populations. *Proc. roy. Soc.* (B), 145, 292 (as for *op. cit.*)

McConnell, R. A. 1961. The Absolute Weapon. *Amer. Inst. biol. Sci. Bull.*, 11, 14. (Eutelegenesis; importance of individuals of high ability.)

Medawar, P. B. 1960 *The Future of Man*: The Reith Lectures. Methuen. London.

Montague, A. 1957. *Anthropology and Human Nature*. Sargent. Boston. (Environmentalist views on race.)

Motulsky, A. G. 1960. Metabolic Polymorphisms and the Role of Infectious

Diseases in Human Evolution. Human Biol. 32, 29. (Morphisms and disease-resistance.)

Muller, H. J. 1929. The Method of Evolution. *Sci. Monthly*, Dec. 1929, 481. (Natural selection and directional evolution.)

—— 1936. *Out of the Night.* Gollancz. London. (Eugenics and eutelegenesis.)

—— 1949. The Darwinian and Modern Conceptions of Natural Selection. *Proc. Amer. Philos. Soc.*, 93, 459. (Natural selection; fitness.)

—— 1950a. Our Load of Mutations. *Amer. J. hum. Genet.* 2, 111. (Genetic load).

—— 1950b. Radiation damage to the Genetic Material. *Amer. Scientist*, 38, 3. (Effects of X rays, fall-out, etc.)

—— 1959. The Guidance of Human Evolution. *Persp. Biol. Med.*, 3, 1 (Eutelegenesis, deep-frozen sperm.)

—— 1961a. (Review of P. B. Medawar's *The Future of Man.*) *Persp. Biol. Med.*, 4, 377.

—— 1961b. Human Evolution by Voluntary Choice of Germ-plasm. Science, 134, 643. (Eutelegenesis, agapogeny.)

—— 1961c. The Human Future: *The Humanist Frame*, ed. J. S. Huxley. London and New York. (Cultural and genetic evolution.)

Pettenkofer, H. J. et. al. 1962. *Nature*, 193, 445. (Blood-group morphisms and disease-resistance.)

Roberts, J. A. Fraser. 1959. *An Introduction to Medical Genetics*, 2nd Ed. O.U.P. Oxford.

Sauvy, A. 1961. *Fertility and Survival*. Chatto and Windus. London. (Overpopulation and economic development.)

Sheldon, W. H. 1940. *The Varieties of Human Physique*. Harpers. New York.

Sheppard, P. M. 1958. *Natural Selection and Heredity*, Hutchinson. London. (Natural selection.)

Tax, S. (Ed.). 1960. Articles by Kroeber, Washburn and Howell, Adams, Steward, *et al.* in *The Evolution of Man*, Univ. of Chicago Press, Chicago. (Evolution and culture.)

Thoday, J. M. 1953. Components of Fitness. *Symp. Soc. exp. Biol.* (Evolution) 96. (Fitness, stability, selection.)

Waddington, C. H. 1957. *The Strategy of the Genes*. Allen and Unwin. London. (Genetic assimilation; genetics and development.)

Williams, R. 1960. Why Human Genetics? *J. Hered*, 51, 91. (Human genetic variance and environment.)

15 Sir Julian Huxley, FRS – Chronological Table

Taken from *Biographical Memoirs of Fellows of the Royal Society*, volume 22, London, 1976.

A few of the dates mentioned in Sir Julian Huxley's *Memories* and in *Who's Who* are inaccurate. They have been corrected in this table.

1887	Julian Sorell Huxley was born in London on 22 June.
1897 or 1898	Entered Hillside Preparatory School, Godalming, Surrey as day-boy.
1900	Entered Eton College with scholarship.
1906	To Germany to learn the language. Entered Balliol College, Oxford, with scholarship.
1908	Awarded the Newdigate Prize for English Verse at Oxford.
1909	First Class in Natural Science (Zoology) at Oxford University. Attended international gathering at Cambridge held to celebrate the half-centenary of publication of *The Origin of Species*.
1909–10	At Naples Zoological Station, as Oxford Naples Scholar.
1910	Appointed Lecturer at Balliol College and Demonstrator in the Department of Zoology and Comparative Anatomy at Oxford.
1912	Studied courtship of *Podiceps cristatus* with his brother, Trevenen, at Tring, Hertfordshire. Appointed Research Associate in Biology at Rice Institute, Houston, Texas, and went there for formal opening of the institute.
1913	To Germany, to work with Otto Warburg and Richard Hertwig. To Rice Institute as Assistant Professor of Biology.
1914	Mild depression. Returned to England.
1914–15	At Rice Institute as Professor of Biology.
1916	Worked at Marine Biological Laboratory, Woods Hole, Massachusetts. Returned to England to participate in the war. Met Miss Juliette Baillot at Garsington Manor, near Oxford.
1917	Enlisted in the Army Service Corps.
1918	At Padua, Italy, as Lieutenant in Army Intelligence. After the armistice he was an Education Officer in the Army.
1919	Appointed Fellow of New College and Senior Demonstrator in the Department of Zoology and Comparative Anatomy at Oxford. Married Miss Juliette Baillot.
1921	Was a member of the Oxford University Expedition to Spitzbergen.
1925	Appointed Professor of Zoology at King's College, London University.
1926	Accepted invitation by H. G. Wells to collaborate with him and G. P. Wells in writing *The Science of Life*.

1927   Resigned Chair of Zoology at King's College to work on *The Science of Life*; remained at the college as Honorary Lecturer in Experimental Zoology. Became an original member of the Birth Control Investigation Committee.

1927–31 Lectured at the Royal Institution, London, as Fullerian Professor of Physiology.

1929   Visited East and Central Africa at invitation of the Colonial Office's Committee on Education.

1930   President of the Association of Scientific Workers.

1931   Visited USSR at invitation of Intourist.

1932–34 Engaged in writing three books.

1934   Collaborated with Ronald Lockley and a professional cinematographer at an island off the Pembrokeshire coast in making the film entitled *The Private Life of the Gannet*. Studied regeneration in *Sabella* (Polychaeta) at Loch Ine, south-west Ireland.

1935   Appointed Secretary of the Zoological Society of London.

1936   Galton Lecture of Eugenics Society: 'Eugenics and Society'.

1936–41 Engaged in writing *Evolution, the Modern Synthesis*, while Secretary of the Zoological Society. Took part in the foundation of the Population Investigation Committee and the Association for the Study of Systematics.

1938   Elected Fellow of the Royal Society.

1939   Visited St Kilda (beyond Outer Hebrides, Scotland) with other naturalists. The BBC 'Brains Trust' was started, with Huxley as a regular member of the team.

1941–42 Lectured in the United States under auspices of Rockefeller Foundation. Visited Tennessee Valley.

1942   Resigned his position as secretary of the Zoological Society. Took possession of 31 Pond Street, London, his home for the rest of his life.

1943   Busy with broadcasting and editing. Delivered Romanes Lecture at Oxford on 'Evolutionary Ethics'.

1944   To West Africa as member of Commission on Higher Education in British Colonies. Afterwards participated actively in movement to include science in the name of the proposed UNESCO (United Nations Educational and Cultural Organization).

1945   To USSR to take part in Bicentenary of the Academy of Sciences.

1946   Appointed first Director-General of UNESCO.

1947   Visited Haiti and fifteen nations of Latin America on behalf of UNESCO. Afterwards to Mexico for General Conference of UNESCO.

1947–48 After the conference visited other countries of Central America.

1948   Visited countries of Near East, northern Africa and Europe on behalf of UNESCO. Decision was taken by UNESCO to establish IUCN (International Union for the Conservation of Nature). Huxley visited Poland (not under auspices of UNESCO) with a party largely composed of left-wing sympathisers. To Beirut for General Conference of UNESCO. Term of office as Director-General of UNESCO ended.

1949   Helped to found Ecological Society and Society for the Study of Animal Behaviour. To Iceland, to study conservation of wild life.

1950   Lectured at Swedish Academy of Science on 'Bird Courtship and Display'. Short lecture tour in the United States.

1951   Lectured at Münster University at invitation of Professor B. Rensch. Lecture tour in the United States. Participated in foundation of the Society for the Study of Evolution.

1951–52 Nervous breakdown.

1953   Received Kalinga Prize for popularisation of science.

1953–54 Visited the United States, Pacific Islands, Australia, Tasmania, East Indies, South-East Asia, India, Iraq, Iran (where he lectured at the millenary celebrations of Avicenna's birth), Syria and Lebanon.

1955   Lectured on cancer in New York.

1956   Made further study of literature of cancer at Woods Hole, Massachusetts. Received Darwin Medal of the Royal Society. Studied possibility of nature reserve at Coto Doñana in southern Spain, with others. Nervous breakdown.

1958   Knighted.

1959   Spoke at Conference on Planned Parenthood in Delhi. Received Lasker Award for Contributions to Planned Parenthood in New York. Spoke at Darwin Centenary in Chicago.

1959–62 President of the Eugenics Society.

1960   To southern and eastern Africa, to report to UNESCO on conservation of wild life.

1961   To western Africa; lectured at Achimota University, Ghana.

1962   Galton Lecture of Eugenics Society: 'Eugenics in Evolutionary Perspective'.

1963   To Jordan with others, to study conservation. Attended IUCN Conference, Nairobi. To Ethiopia to report on possibility of national parks.

1965   Organised Royal Society Discussion Meeting on Ritualisation of Behaviour in Animals and Man.

1966   Nervous breakdown.

1967   Holiday in Tunisia. BBC produced a television programme to celebrate the eightieth anniversary of his birth.

1969   The Golden Wedding of Sir Julian and Lady Huxley was celebrated at the Fellows' Restaurant in the Zoological Gardens, London.

1970   At an International Congress of IUCN and the World Wildlife Fund, Sir Julian received a gold medal and gold watch for 'his outstanding contribution to scientific research relating to conservation'.

1971   To Paris, to attend the twenty-fifth anniversary of the foundation of UNESCO. Revisited national parks in the United Republic of Tanzania.

1972   Visited the Department of Zoology, Oxford University, to unveil portraits of three former members of the academic staff.

1975   Sir Julian died on 14 February. There was a Memorial Meeting at St John's, Smith Square, London, on 18 April.

16 Sir Julian Huxley, FRS – Select Bibliography

Taken from *Biographical Memoirs of Fellows of the Royal Society*, Volume 22, London, 1976. Compiled by J. R. Baker, FRS.

(1) 1912   Some phenomena of regeneration in *Sycon*; with a note on the structure of its collar-cells. *Phil. Trans. R. Soc. Lond.*, B **202**, 165–189.

(2)   A first account of the courtship of the redshank (*Totanus calidris* Linn.). *Proc. Zool. Soc. Lond.*, **2**, 647–655.

(3)   A 'disharmony' in the reproductive habits of the wild duck (*Anas boschas* L.). *Biol. Zbl.*, **32**, 621–623.

(4)   The great crested grebe and the idea of secondary sexual characters. *Science*, **36**, 601–602.

(5)   *The individual in the animal kingdom.* Cambridge University Press.

(6) 1914   The courtship habits of the great crested grebe (*Podiceps cristatus*); with an addition to the theory of sexual selection. *Proc. Zool. Soc. Lond.*, **2**, 491–562.

(7) 1916   Bird watching and biological science. Some observations on the study of courtship in birds. *Auk*, **33**, 142–161.

(8) 1919   Some points in the sexual habits of the little grebe, with a note on the occurrence of vocal duets in birds. *Br. Birds*, **13**, 155–158.

(9) 1920   Metamorphosis of axolotl caused by thyroid feeding. *Nature, Lond.*, **104**, 435 (only).

(10) 1921   Further studies on restitution bodies and free tissue-culture in *Sycon. Quart. F. micr. Sci.*, **65**, 293–322.

(11)   Studies in dedifferentiation. II. Dedifferentiation and resorption in *Perophora. Quart. F. micr. Sci.*, **65**, 643–697.

(12) 1922   (With L. T. HOGBEN) Experiments on amphibian metamorphosis and pigment responses in relation to internal secretions. *Proc. R. Soc.* B, **93**, 36–53.

(13)   Ductless glands and development. Amphibian metamorphosis considered as consecutive dimorphism, controlled by glands of internal secretion. *F. Hered.*, **13**, 349–358.

(14) 1923   Ductless glands and development. II. Amphibian metamorphosis considered as consecutive dimorphism, controlled by the glands of internal secretion. *F. Hered.*, **14**, 3–11.

(15)   (With G. R. DE BEER) Studies in dedifferentiation. IV. Resorption and differential inhibition in *Obelia and Campanularia. Quart. F. micr. Sci.*, **67**, 473–495.

244 *Select Bibliography*

(16)   Courtship activities in the red-throated diver (*Colymbus stel-latus* Pontopp.); together with a discussion of the evolution of courtship in birds. *F. Linn. Soc. Zool.*, **35**, 253–292.
(17)   *Essays of a biologist.* London: Chatto & Windus.
(18) 1924   (With G. R. DE BEER) Studies in dedifferentiation. 5. Dedifferentiation and reduction in *Aurelia. Quart. F. micr. Sci.*, **68**, 471–479.
(19)   (With P. D. F. MURRAY) A note on the reactions of chick chorio-allantois to grafting. *Anat. Rec.*, **28**, 385–388.
(20)   Early embryonic differentiation. *Nature, Lond.*, **113**, 276–278.
(21)   The variation in the width of the abdomen in immature fiddler crabs considered in relation to its relative growth-rate. *Amer. Nat.*, **58**, 468–475.
(22)   Constant differential growth-ratios and their significance. *Nature, Lond.*, **114**, 895–896.
(23) 1925   Studies on amphibian metamorphosis. II. *Proc. R. Soc.*, B, **98**, 113–146.
(24)   (With E. B. FORD) Mendelian genes and rates of development. *Nature, Lond.*, **116**, 861–863.
(25) 1926   Studies in dedifferentiation. VI. Reduction phenomena in *Clavellina lepadiformis. Pubbl. Staz. zool. Napoli*, **7**, 1–35.
(26)   Modification of development by means of temperature gradients. (Record of reading of paper only.) *Anat. Rec.*, **34**, 100.
(27)   The annual increment of the antlers of the red deer (*Cervus elaphus*). *Proc. Zool. Soc. Lond.*, **2**, 1021–1035.
(28)   *The stream of life.* London: Watts.
(29)   *Essays in popular science.* London: Chatto & Windus.
(30) 1927   Further work on heterogenic growth. *Biol. Zbl.*, **47**, 151–163.
(31)   Discontinuous variation and heterogony in *Forficula. F. Genet.*, **17**, 309–327.
(32)   On the relation between egg-weight and body-weight in birds. *F. Linn. Soc. Zool.*, **36**, 457–466.
(33)   Studies on heterogonic growth (IV). The bimodal cephalic horn of *Xylotrupes gideon. F. Genet.*, **18**, 45–53.
(34)   The modification of development by means of temperature gradients. *Arch. EntwMech. Org.*, **112**, 480–516.
(35)   (With E. B. FORD) Mendelian genes and rates of development in *Gammarus chevreuxi. Brit. F. exp. Biol.*, **5**, 112–134.
(36)   Introduction, in E. Selous, *Realities of Bird Life* (London: Constable).
(37)   *Religious without revelation.* London: Benn.
(38) 1928   Sexual differences of linkage in *Gammarus chevreuxi. F. Genet.*, **20**, 145–156.
(39) 1929–30 (With H. G. & G. P. WELLS) *The science of life: a summary of contemporary knowledge about life and its possibilities.* (In 31 fortnightly parts.) London: Amalgamated Press.
(40) 1930   Spemanns 'Organisator' und Childs Theorie der axialen Gradienten. *Naturwissenschaften*, **18**, 265 (only).

(41)   *Bird-watching and bird behaviour*. London: Chatto & Windus.
(42)   *Ants*. London: Benn.
(43)   1931   The relative size of antlers in deer. *Proc. Zool. Soc. Lond.*, **2**, 819–864.
(44)   Notes on differential growth. *Amer. Nat.*, **65**, 289–315.
(45)   (With O. W. RICHARDS) Relative growth of the abdomen and the carapace of the shore-crab *Carcinus maenas*. *F. Mar. Biol. Ass. U.K.*, **17**, 1001–1015.
(46)   Relative growth of mandibles in stag-beetles (Lucanidae). *F. Linn. Soc. Zool.*, **37**, 675–703.
(47)   (With H. G. WELLS & G. P. WELLS) *The science of life*. (Edition in book form; see (39).) London: Cassell.
(48)   *Africa view*. London: Chatto & Windus.
(49)   1932   (With A. WOLSKY) Structure of normal and mutant eyes in *Gammarus chevreuxi*. *Nature, Lond.*, **129**, 242–243.
(50)   *Problems of relative growth*. London: Methuen.
(51)   *What dare I think? The challenge of modern science to human action and belief*. London: Chatto & Windus.
(52)   *A scientist among the Soviets*. London: Chatto & Windus.
(53)   *The captive shrew and other poems of a biologist*. Oxford: Basil Blackwell.
(54)   n.d.   (?1934) (With E. N. DA C. ANDRADE) *Simple Science*. Oxford: Blackwell.
(55)   1934   (With G. R. DE BEER) *The elements of experimental embryology*. Cambridge University Press.
(56)   1935   *Ants* (illustrated edition). London: Chatto & Windus.
(57)   (With A. C. HADDON) *We Europeans, a survey of 'racial' problems*. London: Cape.
(58)   1936   Natural selection and evolutionary progress. *Rep. Brit. Ass.*, **106**, 81–100.
(59)   *At the Zoo*. London: Allen & Unwin.
(60)   *Eugenics and Society*. (The Galton Lecture for 1936) *Eugenics Review*, **28**, 11–31.
(61)   1938   Clines: an auxiliary taxonomic principle. *Nature, Lond.*, **142**, 219–220.
(62)   Species formation and geographical isolation. *Proc. Linn. Soc.*, **150**, 253–264.
(63)   Threat and warning colouration in birds with a general discussion of the biological functions of colour. *Proc. VIII Int. Ornith. Congr.*, 430–455.
(64)   Darwin's theory of sexual selection and the data subsumed by it, in the light of recent research. *Amer. Nat.*, **72**, 416–433.
(65)   (With H. G. & G. P. WELLS) *The science of life* (popular edition). London: Cassell.
(66)   1939   (No title printed; contribution to a discussion on subspecies and varieties, arranged at the request of the Association for the Study of Systematics.) *Proc. Linn. Soc.* **151**, 105–114.
(67)   Clines: an auxiliary method in taxonomy. *Bijdr. Dierk.*, **27**, 491–520.

(68)   (With L. KOCH) *Animal language.* London: Country Life.
(69)   *The living thoughts of Darwin, presented by Julian Huxley.*
             London: Cassell.
(70) 1940   (No title printed; contribution to a discussion on phylogeny
             and taxonomy.) *Proc. Linn. Soc.*, **152**, 251–252.
(71)   *The new systematics.* Edited by Julian Huxley, and with an
             'Introductory: towards the new systematics', contributed
             by him. Oxford: Clarendon Press.
(72) 1941   Evolutionary genetics. *Proc. VII Int. Genet. Congr.*, 157–164.
(73)   *The uniqueness of man.* London: Chatto & Windus.
(74) 1942   *Evolution, the modern synthesis.* London: Allen & Unwin.
(75) 1943   *TVA adventure in planning.* Cheam (Surrey): The Architec-
             tural Press.
(76) 1944   *On living in a revolution.* London: Chatto & Windus.
(77) 1946   *UNESCO, its purpose and its philosophy.* Published by the
             Preparatory Commission of the United Nations Educa-
             tional, Scientific and Cultural Organization. (Place of pu-
             blication not stated.)
(78) 1947   (With T. H. HUXLEY) *Evolution and ethics 1893–1943.* Lon-
             don: Pilot Press.
(79) 1949   *Heredity, east and west.* New York: Schuman.
(80) 1953   *Evolution in action.* (Based on the Patten Foundation Lec-
             tures delivered at Indiana University in 1951.) London:
             Chatto & Windus.
(81) 1954   Article on 'The evolutionary process' in *Evolution as a pro-
             cess.* Edited by Julian Huxley, A. C. Hardy, and E. B. Ford.
             London: Allen & Unwin.
(82)   *From an antique land: ancient and modern in the Middle East.*
             London: Parrish.
(83) 1955   Morphism and evolution. *Heredity*, **9**, 1–52.
(84) 1956   *Kingdom of the beasts.* With 175 photogravure plates by W.
             Suschitzky. London: Thames & Hudson.
(85) 1957   The three types of evolutionary process. *Nature, Lond.*, **180**,
             454–455.
(86)   *New bottles for new wine: essays.* London: Chatto & Windus.
(87) 1958   *The story of evolution: the wonderful world of life.* London:
             Rathbone Books.
(88)   *Biological aspects of cancer.* London: Allen & Unwin.
(89) 1959   Introduction, in P. Teilhard de Chardin, *The Phenomenon of
             Man* (London: Collins).
(90) 1961   *The conservation of wild life and natural habitats in Central
             and East Africa.* Paris: Unesco.
(91) 1962   *Education and the humanist revolution.* (The ninth Fawley
             Foundation Lecture; published by the University of South-
             ampton.)
(92)   *Eugenics in Evolutionary Perspective.* (The Galton Lecture
             for 1962) *Eugenics Review* **54**, 123–141.
(93) 1963   *Evolution, the modern synthesis.* Second edition, with new
             Introduction. London: Allen & Unwin.

(94) 1964   Essays of a humanist. London: Chatto & Windus.
(95) 1965   (With H. B. D. KETTLEWELL) *Charles Darwin and his world.*
              London: Thames & Hudson.
(96) 1970   *Memories.* London: Allen & Unwin.
(97) 1973   *Memories II.* London: Allen & Unwin.
(98) 1974   *Evolution, the modern synthesis.* Third edition, with new
              Introduction by nine contributors. London: Allen &
              Unwin.

Index